I0889268

CONTESTING AGRICULTURE

SUNY series in the Anthropology of Work
June C. Nash, editor

CONTESTING AGRICULTURE

Cooperativism and Privatization in the New Eastern Germany

Hans C. Buechler
and
Judith-Maria Buechler

State University of New York Press

Published by
State University of New York Press, Albany

© 2002 State University of New York

All rights reserved

Printed in the United States of America

No part of this book may be used or reproduced in any manner whatsoever without written permission. No part of this book may be stored in a retrieval system or transmitted in any form or by any means including electronic, electrostatic, magnetic tape, mechanical, photocopying, recording, or otherwise without the prior permission in writing of the publisher.

For information, address State University of New York Press,
90 State Street, Suite 700, Albany, NY 12207

Production by Kelli Williams
Marketing by Anne M. Valentine

Library of Congress Cataloging-in-Publication Data

Buechler, Hans C.
Contesting agriculture : cooperativism and privatization in the new Eastern Germany / Hans Buechler and Judith-Maria Buechler.
p. cm. — (SUNY series in the anthropology of work)
Includes index.
ISBN 0-7914-5281-6 (alk. paper) — ISBN 0-7914-5282-4 (pbk. : alk. paper)
1. Agriculture, Cooperative—Germany (East) 2. Privatization—Germany (East) I. Buechler, Judith-Maria. II. Title. III. Series

HD1491.G32 E273 2002
334′.683′09431—dc21

2001049286

10 9 8 7 6 5 4 3 2 1

CONTENTS

ILLUSTRATIONS

Maps

Figures

Tables

ACKNOWLEDGMENTS

Writing this book would not have been possible without the assistance of many individuals. The book had its inception during a short visit to eastern Germany in 1991, Judith-Maria's first to Dessau, her parents' place of origin that they had left as refugees in 1932. We were intrigued by her aunt's (now deceased) description of life in the GDR and her house whose furnishings and kitchen reminded Hans of houses he had visited as a small child in Switzerland just after World War II. We were fascinated by the small businesses that had sprung up like mushrooms in the middle of the residential neighborhood, enterprises that included a store specializing in selling surfboards at least six hours by train from the closest ocean. Judith-Maria's cousin, whom she had last seen as a teenager in Berlin in 1961, now had sons who told us about the excitement but also dislocation they had experienced right after the fall of the Berlin Wall. After having undertaken research during the debt crisis in the early 1980s in Bolivia as well as the transition from dictatorship to democracy in Spain, research on the turmoil surrounding the transition from communism to capitalism seemed like a logical next step in our pursuit of understanding adaptations to rapid change. We thank Hildegunde, Arno, Benjamin and Bernhard Sames for introducing us to this new research topic and later for hosting us at their home in Halle. In addition, we wish to thank Judith-Maria's cousin Renate Rarichs and her husband Karl for helping us launch our fieldwork by hosting us in Frankfurt and assisting us in purchasing a secondhand car.

Field research in Germany was supported by a Fulbright grant, a grant from the Deutscher Akademischer Austauschdienst (DAAD), and sabbatical pay from Hobart and William Smith Colleges and Syracuse University. In Germany, Prof. Michael Tolksdorf of the Fachhochschule für Wirtschaft Berlin and Prof. Rudi Schmidt at the Institut für Soziologie

Friederich-Schiller-Universität Jena graciously acted as local sponsors for our research. Many thanks to both.

Most important, we wish to thank our consultants in Sachsen-Anhalt for their patient explanation of their history and their present lives. In particular, the farmers interviewed for this study granted us interviews of such length and detail that we reoriented our research—which was initially focused on urban occupations—to include farming as well.

Finally, we wish to thank June Nash and the anonymous reviewers of State University of New York Press for their invaluable comments. Parts of this book were published in two articles, one in *Identities* (Buechler & Buechler 2000), reprinted here with permission from Gordon and Breach Publishers, and one in the *Anthropology of Work Review* (Buechler & Buechler 1995). We wish to thank Nina Glick and the anonymous reviewers of the former article for their input. However, responsibility for any errors and omissions is ours alone.

ABBREVIATIONS

ABM	*Arbeitsbeschaffungsmaßnahmen*: workfare programs to combat unemployment
BDB	*Bund der Deutschen Bauern*: Union of German Farmers
BVVG	*Boden-Verwertungs-und-Verwaltungs-Gesellschaft*: agency in charge of privatizing land (successor to *Treuhand*)
CDU	Christian Democratic Union (one of the major German political parties)
EC	European Community
eG	*eingeschriebene Genossenschaft*: legally constituted cooperative
EU	European Union
GATT	General Agreement on Tariffs and Trade
GbR	*Gesellschaft bürgerlichen Rechts*: commonlaw partnership
GDR	German Democratic Republic
GmbH	*Gesellschaft mit beschränkter Haftung*: limited enterprise
LPG	*landwirtschaftliche Produktionsgenossenschaft*: agricultural collective
MTS	*Maschinen-Traktoren Stationen*: farm machinery pools
PDS	political party that succeeded the Communist Party after the *Wende*
SED	Eastern German Communist Party

SPD Social Democratic Party

VBE *Vollbeschätigte Einheit:* unit of full-time work

VdgB *Vereinigung der gegenseitigen Bauernhilfe*: farmers' mutual aid association

VeB *volkeigener Betrieb:* state-owned enterprise

CHRONOLOGY

1849–1862	Founding of the first cooperatives to secure raw materials and credit	
1867	Institution of the Prussian cooperative law	
1871	Cooperative law extended to the entire German *Reich*	
1945	Russian-instituted agrarian reform (*Bodenreform*)	
1949	Founding of the GDR	
1952	Pressure is initiated toward collectivization and centralization of decision-making in agriculture	
1960	Forced collectivization	
early 1970s	Consolidation into large collectives	
1978	Separation of crop and livestock collectives	
1989	Nov. 9	Fall of the Berlin Wall
1990	March 1	Founding of the Treuhandanstalt (privatization agency)
	June 29	Law governing the transformation of LPGs (*Landwirtschaftsanpassungsgesetz)* enacted
	July 1	Currency Union
	Oct. 3	Reunification of the two Germanies
1992	Jan. 7	Special agency for the privatization of 1.5 million hectares of state land created (*Bodenverwertungs-und-verwaltungs* GmbH [BVVG])

1996	Jan. 15	Old debt to be reduced through the sale of nonessential assets of a cooperative such as apartments; otherwise, forced sales

CHAPTER ONE

Introduction

The subject of this book is the transformation of farming after the demise of communism in eastern Germany and the manner in which individuals, both eastern Germans and western Germans who migrated there after 1989, conceive of this process and of their roles and identities within it. As such, this book is about contradictions. It is about a farm manager of a highly rationalized, modern cooperative who still has strong beliefs in the basic correctness of the communist path. It is about managers in agricultural cooperatives who preside over the reduction of the workforce to a fraction of 1990 membership in keeping with capitalist notions of efficiency, but who are also firmly committed to preserving as many workplaces following an ideology that harkens back to socialist times. It is about a newly created eastern German millionaire individual farmer who seemed to live in poverty. And it is about a western German farmer, a newcomer to the area, who is the beneficiary of generous subsidies from the federal government and from the *Land* (federal state), but whose farm is subject to potentially disastrous inundation, a problem that had not unduly affected the former German Federal Republic (GDR) collective because it constituted only a small part of the extensive territory farmed.

The *Wende* (literally, the turn or turning), the fall of the Berlin Wall in November 1989 and the reunification of the two Germanies in 1990, was a major watershed in eastern Germany. Not only did it mean the demise of communism, an event shared with a number of other eastern European countries, but it meant the country's absorption into what had become a foreign nation. We shall follow Darnton's (1991:21–26; see also Bornemann 1991) chronology of the events. The *Wende* was preceded by a tumultuous year that began with Hungary's decision to begin to dismantle its border with Austria in May 1989, whereupon eastern Germans began to flock to Hungary and seek asylum in the West German embassies in Prague and Warsaw. By October, 55,000 East Germans had fled to the West and an additional 17,000 traveled from Prague to West Germany in special trains. After a series of mass demonstrations in East German cities, Erich Honecker, the general secretary of the Communist Party, was replaced. But, after new demonstrations and massive emigration via Hungary and Czechoslovakia, the Berlin Wall that divided the city of Berlin was opened on November 9 and travel to West Germany was permitted freely. A new government, under Hans Modrow, began to institute a series of reforms, including the elimination of the leading role of the Communist Party; the abolition of the dreaded *Stasi*, the secret police; and the initiation of privatization of state enterprises as well as the restructuring of collectives. These reforms were catalyzed by round tables of representatives of citizen's movements and political parties, who, for a while, attempted to create a "third way" between communism and Western democracies. These developments led to free parliamentary elections and the replacement of Modrow by de Mazière of the Christian Democratic Union (CDU), the party in power in West Germany. These elections and the pressure of West German politicians, including negotiations between the West German Chancellor, Helmut Kohl, and Gorbachev also led to the rapid push toward reunification. On July 1, 1990, monetary union took effect and on October 3 the five states of East Germany were incorporated into the Federal Republic of Germany.

The clash between the two radically different systems resulted in major contradictions in the eastern German economy, in general, and in agriculture, in particular. Indeed, agriculture in the GDR was fully collectivized, with the exception of a handful of small farms. Members of farming collectives were permitted to cultivate small plots of land and raise a few animals, and they still held title to (but not control over) the land they had brought into the collective. In Poland, in contrast, a large part of farming remained in private hands throughout the communist period and in certain other countries in the Soviet bloc, particularly toward the end of communism, there were all sorts of arrangements between the collectives and individual farmers giving them more autonomy (see chapter 4). The economic system introduced in eastern Germany after the *Wende* was in some ways even more capitalistic than the version prevailing in western Germany. It thus comes as no surprise that forms with inconsistent characteristics would initially emerge.

It is our contention, however, that a model of change that presents the process as the transformation of one pure or coherent form into another with transitional intermediate forms, or even a model that depicts change as resulting in a more permanent amalgamation of two coherent antecedent forms, is too simplistic. Socioeconomic systems can rarely if ever be represented by unitary models without doing violence to their actual dynamics. Thus, both GDR socialism and western German capitalism entailed many often conflicting currents resulting from the persistence of a variety of cultural traditions, the outcomes of class conflict, the need to remedy the inadequacies of each system by allowing workable solutions even if they were antithetical to the central tenets of the hegemonic system, the pressures of national and international politics, and global economic linkages.

An analysis of the transformation of agriculture in eastern Germany must, then, address the conflicting trends during different historical periods that, in turn, form the basis of central or of countervailing trends in subsequent periods. In the area studied, the central region of the state, or *Land,* of Sachsen-Anhalt, modern forms of agriculture were influenced by the area's high fertility but low rainfall. The precommunist preponderance of large-scale farming dominated by estates in the hands of nobility has also left its mark and long-lost patterns of ownership are again becoming relevant. The communist past is manifested less in echoes of the then dominant form of production, the centralized, state-owned enterprise (*volkseigener Betrieb*), which never became prevalent in agriculture, but rather in the continued influence of a compromise form, the collective. But modern forms of agriculture are not only inspired by these long-lasting historical forms, but by the specific history of collectivism in the GDR. The fact that the large precommunist estates were first divided into small individual plots and given to farmers in the region and to refugees from the east, that these individual holdings were then grouped into cooperatives with various degrees of joint production, and that these cooperatives were in turn grouped into progressively larger entities have all influenced postcommunist farming in eastern Germany. Finally, the survival of individual production on half hectare plots assigned to each farm worker household, although integrated into the communist redistributive economy, also has had its impact on present-day agriculture.

The present situation is one of experimentation with both old and new forms within the confines of wider agricultural policies that are themselves in flux. Thus, in the mid-1990s, and, to a lesser extent, even in 1999, the rules of privatization developed by the German authorities continued to be ambiguous: they vacillated between attempts to destroy forms dominant during communism, including those, like cooperatives, which have a much longer history in Germany, and policies (first developed during the transitional period of the Modrow regime, early in 1990) to authorize any form of enterprise that respected the basic underpinnings of capitalism. The result has been the emergence of three major actors: producer cooperatives, erstwhile members of agricultural collectives who established their own farms, and western individuals and corporations who gained a foothold in the east.

As we shall see, the history of the cooperatives is characterized by a progressive curtailing of the autonomy of individual farmers during the four decades of communist rule followed by new modes of centralized control. The first cooperatives, instituted in the 1950s, were based on the nineteenth-century model of Raiffeisen.[1] They were characterized by voluntary association into various forms of cooperatives with differing degrees of pooling of land and livestock. Subsequently the voluntary nature of this association was curtailed, first by employing economic incentives and disincentives and ultimately through various forms of intimidation. While land titles remained in the hand of individuals, land ownership no longer provided any financial benefits. Nevertheless, remuneration continued to depend, to a degree, on the productivity of the enterprise. Former managers of collectives thus stress that, unlike their counterparts in state enterprises, they enjoyed a degree of agency: the possibility of influencing the living conditions of collectivities smaller than the nation as a whole. Hence they saw their roles as, to a degree, counter-hegemonic. After the *Wende*, producer cooperatives were seen by the west as not fully compatible with capitalism. Many western experts argued that decision-making in producer cooperatives was inherently cumbersome and that the wish to maintain employment, that could be expected to be a major aim of the membership, ran counter to the long-term survival chances of a cooperative. Indeed, managers took upon themselves major economic sacrifices in the form of lower remuneration (albeit with lower personal financial risk) than if they had struck out on their own or had adopted other corporate forms. At the same time, the managers' freedom of action continued to be restricted by state intervention. This time it took the form of decisions regarding debts accrued in GDR times; the obligation to abide by proposals that had to be submitted to the state authorities in order to be able to continue renting land from the state; rules about preferential access to state land; and quotas and subsidies conceded by the federal states, the German government, and the European Union (EU).

A major aim of the German government after the *Wende* was the reinstitution of individual farming following the western German model in the east. Farm families who had joined collectives either voluntarily or by force were particularly encouraged to reestablish farms and were placed first in line in renting state land. Farms that did not exceed a certain level of capitalization also received subsidies in the form of outright grants and low-interest loans.

Finally, western farmers who established farms in the east shared some of these benefits. These farmers included former owners of farms exceeding 100 hectares and therefore expropriated in their entirety during the Soviet land reform of 1949. While not regaining title to the land they lost, they nevertheless received preferential treatment in leasing land that remained in (or had reverted back to) the hands of the state. At the same time, westerners without ties to the region established themselves there. Some contented themselves with renting land and coming over with equipment loaded on flatbed trucks for planting and harvesting, leaving other tasks to local caretakers. Others made permanent moves. Where else could they rent hundreds of hectares of land at advantageous rates made possible through loans secured through ties with bankers and their own or their parents' farming operations back home?

The conflicting trends in the region's past are manifested in the concrete decisions of the present-day actors who, on the one hand compete with one another and apply different models in their farming methods, and, on the other, emulate one another in their attempt to find solutions that are both appropriate to the climatic and landholding structure in the region and adapted to the exigencies of the Common Agricultural Policy of the European Union and global competition. As we shall see, especially during the early transition period after the fall of the Berlin Wall, but to a degree even ten years later, the various actors were differently positioned vis-à-vis global actors in the world market, with the westerners at a clear advantage in terms of knowledge and connections with such actors. Their experimentation has resulted in a multiplicity of forms, some of which may well become dominant in the future while others may disappear or become subject to even further transformation. It has also led to divergent management styles, grounded in past experiences but modified to take advantage of the opportunities offered by the new eastern German situation.

To comprehend these complexities, then, requires models that take into account the competition among forms of agriculture and of their resultant variability as they contradict, collide, mutually reinforce or accommodate, or fuse into unexpected new forms. At the same time, such models must take into account the various ways in which regularities of interpersonal behavior are expressed or contradicted by official pronouncements and regulations and in the everyday words of the actors themselves. An analysis of the nature of the relationship between behavior and symbolization is particularly important in situations of conflict and of rapid change, where it can reveal tensions and resistance against hegemonic forces as well as attempts by individuals to reformulate new social situations in terms of older patterns of action.

In order to tease out the strands of German farmers' identities in terms of informants' self-perceptions—the so-called emic approach—while at the same time grounding these perceptions in changing patterns of behavior influenced by local, regional, and supraregional forces, we shall resort to a theoretical framework that combines more behavior-oriented analyses such as those based on Arensberg's concept of "culture as emergent" (Arensberg 1972, 1981), social network analysis, and political economic approaches with more interpretive approaches. We borrow from Arensberg the notion that in order to specify the nature of continuity and change in specific historical sequences, particularly as they are manifested in small-scale events, we must describe patterns of social behavior in terms of minimal sequences of interaction.[2] He demonstrated that such sequences could emerge de novo as adaptations to new situations, or existing patterns could be applied to new ends. Emerging patterns of social behavior could also be superimposed on existing ones, thereby modifying them into new and more complex patterns.

Since Arensberg gave few specific indications of how one pattern could be transformed into another, he paid little attention to individual agency; and he regarded the ideational and communicative aspects of culture, in the form of symbols and explicit norms, as largely derivative or supportive of basic patterns of interaction. Therefore, his model requires some reformulation. Network models such as that developed by Gulliver (1971) enable us to understand individual contributions and show how interactional sequences that cut across geographic and cultural boundaries influence social systems. In addition, we must include in our model a means of showing how regularities in interpersonal behavior are cognicized, and the ways in which emic models influence interaction.

For such a model to do justice to the complexity of actual social situations, it must explicitly recognize the indeterminacies in both regularities of interpersonal behavior and their symbolic expression. One way in which these indeterminacies in self- and mutual definition in the context of groups with unequal power have been addressed is in the hegemony-resistance model. In particular, Laclau and Mouffe (1985) have highlighted the contingent nature of hegemonies and the multiplicity of struggles (see also Buechler & Buechler 1999a). Their approach stresses the articulatory nature of hegemonies with each attempted articulation transforming both the articulated elements and the hegemony itself. Employing a similar approach, the Comaroffs (1991) show the differences and conflicts between the hegemonizing nonconformist church and the various political actors in South Africa. In this book we largely adopt their approach but give more emphasis to the economic dimensions of struggles for hegemony and hence for identity. Specifically, we shall investigate the impact on the farmers' identities of the uneasy truce between cooperative forms of agriculture and state enterprises under communism and the new configuration of forms that are both more compatible with but also partly antithetical to capitalism.

The new configurations involve a different integration of farms in the wider community and new relationships to other businesses. In the GDR, collectives fulfilled a wide range of economic roles. Inefficiencies in the allocation of resources including services forced firms, both collectives and state enterprises, to attempt to secure a high degree of self-sufficiency through the horizontal integration of economic activities. They undertook their own repairs, including the manufacturing of spare parts, they had their own construction brigades, built and maintained roads, and provided services for the membership ranging from kindergarten and dining facilities to building homes for members and maintaining vacation resorts. Following an analogous model of local self-sufficiency in the absence of adequate allocation of resources from the center, collectives were also cajoled into paying for or building facilities ranging from dance halls to hospitals for the local communities and even district capitals. After the *Wende*, both forms of horizontal integration were sharply curtailed or eliminated altogether with a commensurate loss of employment. The recent history of farming, then, involves a desegregation of economic activities and a greater degree of separation of farms from the wider community. At the same time, given the intense competition for land, farmers and cooperatives now need to woo the owners of parcels of land with invitations to celebrations and New Year's greetings that are designed to make these persons feel more part of the farming community and thus assure the continuation of precious leases. These new managerial strategies impact their definitions of self vis-à-vis other farmers and community members.

The models employed in the analysis of identity have not uniformly acknowledged the flexibility, indeterminacy, and contingent nature of hegemonies. In some studies, for example, *resistance* is depicted as based on flexible, often contradictory conceptualizations while established *hegemonies* continue to be viewed in unitary terms.[3]

A more appropriate model for the analysis of the interaction between part-societies with unequal power is emerging out of research that focuses on the blurring of boundaries between cultures[4] and reflexive analyses of anthropologists in fieldwork situations. We must go beyond the reinterpretation of the hegemonic by subaltern groups to the rethinking of the nature of the boundaries between the hegemonic and the anti-hegemonic and the questioning of the unitary nature of the hegemonic itself. For example, the Comaroffs' work (e.g., 1986, 1991) problematizes the nature of the interface between hegemonic and subaltern groups.[5] This blurring of boundaries is also apparent in the recent reflexive literature of anthropologists in fieldwork situations. Lavie (The Hajj et al., 1994), Narayan (1993), Visweswaran (1994), and Buechler and Buechler (1999a, 2000) all show how the identities of individuals (in these cases anthropologists and their informants), even when they come from very different backgrounds, are also intertwined in various, often unexpected ways. The interrelationship between different types of farmers in Sachsen-Anhalt can be analyzed in similar terms for, depending on the context, these farmers identify with or differentiate themselves from one another and regard themselves as being part of or resisting dominant power structures. Hegemonic power structures themselves are also not unitary. Their inconsistent, equivocal nature permits them to avoid direct clashes with competing systems, but also may facilitate their eventual transformation into or replacement by different hegemonic forms. This lack of uniformity is implicit in many of Gramsci's (1971) original formulations of the concept of hegemony. In anthropology, Nash (1987, 1989) bases her analysis of the role of General Electric in Pittsford on the malleable nature of corporate hegemony.

We shall argue that identities are constructed in the process of confronting the new realities of everyday experience, filtered, and expressed by means of available, but often quite inadequate, cultural categories whose meanings are adapted to new ends. The social identities of our respondents are framed by origin, generation, education, and social class, and by the relationships with other actors. They express themselves in work and life plans and in familial and collective ideologies. The identities are constructed in reaction to a new legal framework, politics and economic interests, and the vagaries of climatic and ecological conditions. All our respondents are trying to build a modern agriculture that is socially responsible, but their definition of responsibility depends on their ideological positions and historical backgrounds. The cultural constructions employed by them are not necessarily consistent, for they retain unresolved contradictions of which the actors are only partially aware.[6]

Recent anthropological scholarship on the transformation of both communist and precommunist socioeconomic structures in eastern Europe has often been sensitive to the contradictory or ambivalent nature of both the original structures and the structures that emerged after the fall of communism. Thus Creed (1998) shows how, during communism, a system ostensibly geared toward industrial production in state- and collective-owned factories and collective farming flexibly accommodated private agricultural production. Thereby, he demonstrates that models based on a categorical opposition between communism and capitalism fail to provide an understanding of the complex adaptations that occurred after 1989. Similarly, Nagengast (1991) has argued that, paradoxically, small- and medium-scale private agriculture flourished in Poland in the last years of communism, but has languished since then. Vasary (1987) and Szelényi (1988) both write about the uneasy coexistence of two economic systems, one socialist-bureaucratic, the other private/entrepreneurial, in Hungary in the 1980s. And, with reference to the same country, Lampland (1995:1) argues that "the process of commodifying labor has been fully realized under socialism in conditions thought to be inimical to capitalist development in general, and to commodification in particular," and, more specifically, "was achieved by the policies and practices of Hungarian socialism" (5). Finally, speaking about the development of social identities in a rural community on the border of the former GDR, after 1989, Berdahl (1999:1, 9) argues that "articulations, ambiguities, and contradictions of identity are especially visible in moments of social upheaval" and that rather than simply accepting the hegemony of the west, "through a dynamic and subtle interplay of imitation and resistance, the inhabitants of this borderland are seeking and asserting new forms of identity."

For the purposes of our model we shall define power relationships underlying hegemony broadly.[7] Rather than adopt a model of power relationships as a monolithic pyramid, the analysis of identities must take into account the multiple, often-conflicting power relationships in a given geographical area. Such an accounting is crucial even in the analysis of seemingly unitary systems such as communist bureaucracies that were material in shaping the past, and to a degree, the present identities of many of the farmers in our study. Creed (1998), for example, has shown how daily practice "domesticated" the communist power structure in Bulgaria, both threatening bureaucratic control but also enabling the state to function by overcoming inherent rigidities. A model of power relationships must also take into account diffuse systems of power, such as those manifested in the global social networks of western capitalist producers and the denser but more spatially circumscribed ones of eastern German farmers.

In addition to contributing to the conversations regarding hegemony and identity, this book will also deal with questions regarding history and memory. The construction of old and new identities both reveals and hides events and thus provides both the informant and the anthropologist with new perceptions of the passage of time and the meaning of history. Our eastern informants distinguished the time after the *Wende*, the fall of the Berlin Wall and reunification from *unsere Zeit,* "our time," or even *bei uns*, "with us," of the GDR, as though they had somehow been displaced not only temporally, but also spatially into a new world. *Unsere Zeit* was conceptualized as an immutable block that contrasted with the period after the *Wende* that seemed timeless in its duration because of the rapidity with which change was taking place. The packing not only of events but also of changes of interpretation into a few years made these years appear as endless. In fact, it was often difficult for our informants to remember events and their impressions of events that had occurred only two years earlier, and when they did, they remembered them as though they had occurred in a distant past, a past only slightly less distant from the present than *unsere Zeit*. Time is manipulated in other ways as well. Both the more distant (that is, precommunist) and the more immediate past are reinterpreted to rationalize present actions and future plans. The new forms of agriculture as institutions emerge out of divergent conceptualizations of points of origin. Rationales for particular policies regarding the formation of farms are based simultaneously on promising survivals of older forms in the shadow of the hegemonic model during particular eras: feudalism, early capitalism, the Soviet occupation, and the GDR or memories of forms that were regarded as having been arbitrarily destroyed by governmental decisions during subsequent hegemonies, culminating in the attempts of both the allied powers and the Kohl government to devise a legal baseline for access to property within a capitalistic framework.

The western German migrants minimized the importance of the forty years of separation and tied their experiences to a common pre–World War II history. They also viewed the time in the east as one of adventure akin to the common German practice of the *Lehr- und Wanderjahre* (time of study and travel) of the apprentice/student. But our informants also had an evolutionary view of time. Thus, GDR agricultural collectivism was seen as having passed through a series of stages with roots in nineteenth-century Raiffeisen cooperatives and developments during the communist period from the tentative and voluntary grouping of small individual holdings that emerged after the land reform of 1949 that expropriated large landed estates, to less than voluntary association and to larger cooperatives that ultimately also became specialized in crop or livestock production. The official version of this developmental vision was a futuristic one according to which a purer form of capitalism would be developed in the east, one that did away with all vestiges of the communist past and at the same time broke with much of the cultural traditions of the old *Bundesländer*.

The Setting

The region where the farms we studied are located lies two hours southwest of Berlin, midway between the cities of Magdeburg and Halle. It is centered on the medium-sized industrial city (40,000 inhabitants) of Bernburg, where two of our independent farmer informants were based in 1994. The others ran their operations from smaller cities and villages within a radius of some 15 kilometers. Nienburg, a locality of 5,000 inhabitants, is typical of small cities in the region. Small as it is, the old city boasts an impressive Romanesque church, a hotel, and a small central square surrounded by stores that have shed their GDR looks of decrepitude. During the Christmas season, it even had a Christmas fair, a feat other localities of this size have tried to emulate unsuccessfully. But the town has been badly hit by the persistent unemployment in the region. According to our informant, an independent farmer, its 27% unemployment rate is the highest in the *Land* (state) of Sachsen-Anhalt. The loss of jobs through the rationalization of such gigantic industries as the soda and cement factories in Bernburg has recently been compounded by the closing of the western-run paper bag factory and the downturn in the construction industry. As for the construction company that went bankrupt, it opened again under another name, but not before leaving a trail of unpaid bills in its wake.

Like many communities in eastern Germany, Nienburg attempted to attract enterprises by establishing an industrial and commercial zone on a nearby 18-hectare open field, but to no avail. The western chain stores that were the most likely candidates had already overextended themselves in the region. As a result of such developments, the population of Nienburg dropped by 9% between 1996 and 1999. In spite of the unemployment and the fact that many did move away from the city, only 40% of the workers of the paper mill that closed took advantage of the company's offer to move to another plant 90 kilometers away. Of those who did, most commute from Nienburg, returning home for the weekend. Our informants told us that eastern Germans did not like to go away from home and avoided a move whenever possible. For example, a manager of a large factory in a medium-sized city told us how traumatic it had been to move from Leipzig, a mere hour and a half away by car or train, and how deracinated he had felt at first in the new surroundings. Another

MAP 1.1. The Federal State of Sachsen-Anhalt and the *Kreis* of Bernburg.

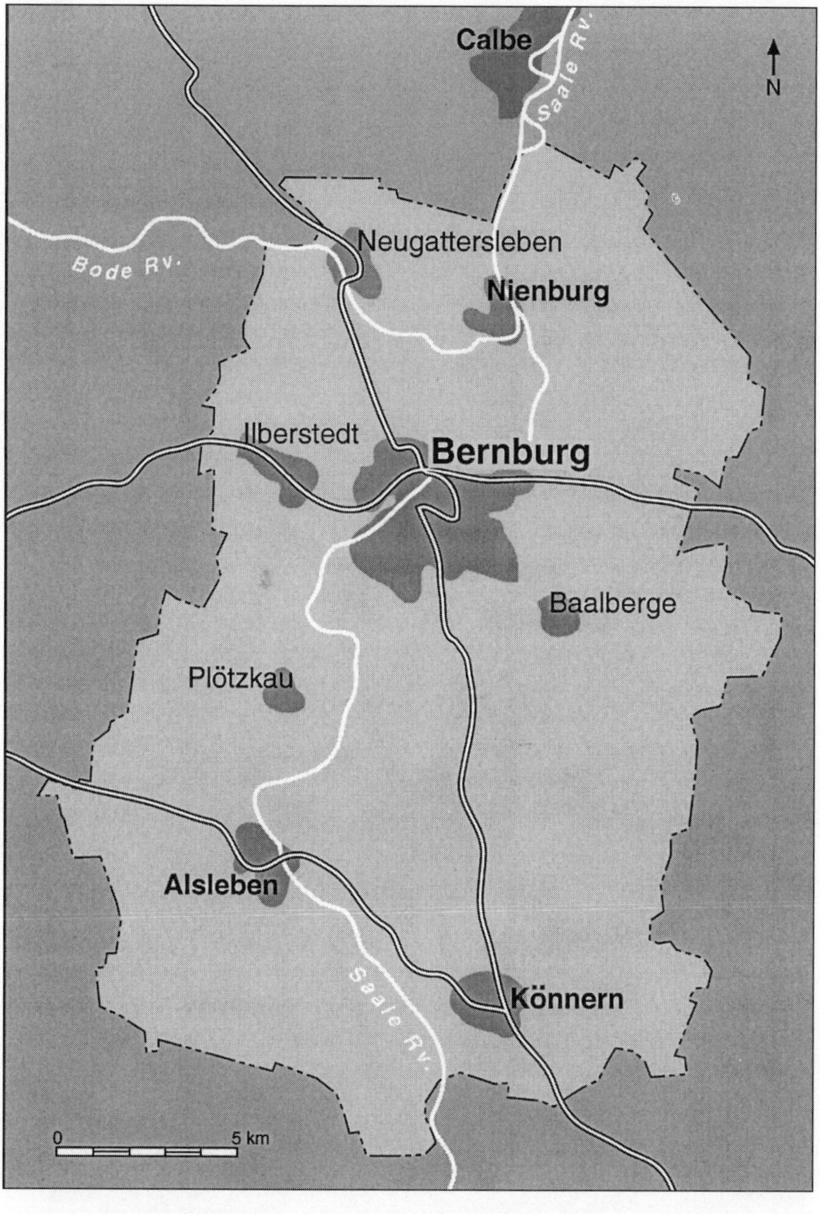

MAP 1.2. *Kreis* Bernburg.

FIGURE 1.1. The Castle of Bernburg.

informant told us jokingly that he felt homesick for Bernburg the moment he drove out of the city. The fact that easterners frequently had to take jobs in the west for a fraction of what their western counterparts were paid only enhanced their reluctance to move. As for the villages around Bernburg, localities with a few hundred inhabitants or even less, many seemed oriented toward larger places. Perhaps because the houses tended to be privately owned in these villages, they appeared to be in better shape than those in the larger localities at the time of the *Wende*. In addition, the communities themselves and individuals have been able to take advantage of state funds to renew roads, public buildings, and private housing. However, even such basic amenities as a grocery store were often lacking, and inhabitants have had to seek services in larger localities, children were bussed to school to nearby Bernburg or to smaller towns, and church congregations were pooled among neighboring settlements.

The farmers' homes, or in the case of the cooperative managers, their offices, were usually located on the outskirts of these settlements, but some lay deep within them. We could not believe our ears when we visited one of our informants who lived a few blocks from the apartment we rented in Bernburg when he said that he was raising horses and pigs in the middle of a city block. While he had moved a few kilometers away by the end of 1999, the barn of another family was still only two blocks away from the center of a smaller city. While most of the individual farmers live next to their barns, a few commute several kilometers to their farms. The cooperatives usually have many of their facilities concentrated in one location, but also have barns and storage facilities dispersed over the landscape, some located within village or hamlet boundaries, others built in open fields. Among the reasons for the dispersal is the fact that present-day cooperatives were often formed out of reshuffled units that were once autonomous and these, in turn, were composed of smaller collectives that were united into larger ones.

The landscape around Bernburg is open and flat or gently rolling with a few low hills here and there, dotted with clusters of trees and fed by streams. The soils are among the richest in the country, but the region lies in the rain shadow of the Harz Mountains, substantially reducing rainfall and therefore making the area exceptionally dry by German standards. While the quantities of crops harvested are affected, the conditions are ideal for growing high-quality wheat. Some of the fields also lie in the floodplains of the storied Saale River and its tributaries, subjecting them to periodic flooding. They are particularly suited for growing hops and vegetables. Even parts of the city of Bernburg are regularly flooded, so the traffic must be detoured over a special flood bridge.

FIGURE 1.2. A village near Bernburg.

Methodology

The basis for this book is a corpus of interviews with some thirty farm owner-managers, farm directors, tractor dealers, agricultural equipment manufacturers, and millers which was part of a larger study of 150 artisans, producers, independent professionals, and retailers. The larger sample will serve to consolidate those aspects of changing identities and underlying transformations that the farmers share with other private entrepreneurs. The initial research took place between September 1993 and May 1994. In November–December 1999, we returned to Bernburg and undertook follow-up interviews with more than half of the farm owner-managers and cooperative managers and one of the millers. Finally, in June and July 2001, we updated our interviews with informants from the other categories of entrepreneurs and professionals in our wider study, including the second miller and the agricultural machinery dealer whose cases are presented in chapter 9.

We chose the region of Bernburg in the *Land* (state) of Sachsen-Anhalt because of the requirements of the wider study. The city of Bernburg fulfilled our criteria of representative size, the presence and mix of industry, commerce, and service, and the rate of unemployment. We also wished to avoid the influence of special circumstances, such as the direct influence of Berlin or a place that was within easy commuting distance of localities in the former Federal Republic of Germany (FRG). Also, Bernburg is located in an area that has been the subject of some studies of an extensive nature but has not been studied intensively with anthropological methods. We only realized after we initiated fieldwork how central agriculture was to the regional economy and the city due to the long tradition of large-scale mixed agriculture. Although our original research plans had focused primarily on urban occupations in the belief that a regional study would be impossible to carry out in the allotted time frame, we soon found that rural and urban were so strongly intertwined in the area that the exclusion of one or the other would have been arbitrary. The fact that we were able to cover all but a few of the farming operations in the *Kreis* (administrative unit) of Bernburg and were even able to add a few beyond its jurisdiction convinced us of the feasibility of expanding our initial focus. This decision was also influenced by the fact that initial interviews revealed that agriculture was one of the few areas of the eastern German economy where major aspects of GDR economic institutions persisted and, in fact, remained dominant. We decided to write a book on this topic first, because it presented a unique opportunity to compare three major different forms of privatized enterprises within the same sector of the economy. Last but not least, the choice was based on the extraordinary willingness of every one of our informants to spend long hours educating us about the complexities of eastern German agriculture, both past and present. We happened to begin our interviews with farmers in December, the beginning of a slack period in agriculture, so we did not feel that we were imposing an overly heavy burden on their time.

When we paid our first visit to eastern Germany in the summer of 1991 to investigate the feasibility of the proposed research, we feared that the experiences of their communist past would make eastern Germans reluctant to talk to strangers. We found the opposite to be the case. People appeared to relish their newly found freedom from secret police surveillance. Talking to newcomers may even have constituted a welcome opportunity to sort out the dizzying changes they were experiencing. Our subsequent fieldwork confirmed this first impression for the most part. The magnitude and the rapidity of the transformation in their lives and their awareness of the uniqueness of the period in which they are living may have contributed to their openness to an opportunity to sort out their ideas in front of strangers with a sympathetic ear.

Our research was facilitated by our fluency in German and the fact that neither researcher is of western German origin, which easterners could have seen as introducing a potential bias. Judith-Maria's parents were born in Sachsen-Anhalt and her great-grandfather was a large farmer in the region. Hans is Swiss. At the same time, the westerners assumed we would understand the difficult position a westerner would experience in a former communist country. Our status as foreigners working in a distant country also lent some prestige to our endeavor, especially after we had been interviewed by the local press.

In our previous research we have found that a female-male team, interviewing together or separately as the situation warrants, greatly facilitates breaking down gender barriers in fieldwork situations and reduces the gender biases on the part of the researchers themselves. A female-male team approach was particularly useful in eastern Germany, since divisions based on gender are quite pronounced. In addition to being a team—since age and professional status are particularly emphasized in Germany—our late middle age and professional status as professors lent credibility to our research and entrée to both younger and older, well-educated farmers. On the other hand, it seemed to matter to our eastern informants that we did not arrive at their doors in a fancy car but in a smallish ten-year-old one and did not put on airs.

Our research in eastern Germany was also facilitated by our own previous research on agriculture, economic change, and industrialization in Spain and Bolivia, which provided us with data for potential comparisons as well as observational and analytical tools. Work in Bolivia and in Spain on various economic issues facilitated an understanding of the various ways in which the economic decisions of individuals are influenced by the various network and institutional contexts in which they operate. Our work led to an appreciation of differences in two widely divergent economies, but also to the range of options *within* each economy and to the similarity of some of these options in the two countries, albeit inserted in divergent ways in the political economies of their respective countries. A holistic approach led us to avoid the adoption of trendy categories that could lead to facile analogizing.

The research followed the anthropological precept of providing a comprehensive portrait of the lives of the farmers. During our first field trip alone, the extraordinary openness of our informants, managers of cooperatives, eastern German, western German, and Dutch individual farmers alike and their willingness to spend as many as six hours in one stretch answering our questions in great detail resulted in a corpus of close to a thousand single-spaced typed pages of transcribed interviews on subjects ranging from the evolution, organizational structure, and operation of collectives during communism; to the transformation of the collectives after unification; adaptation of western farmers to life and agricultural conditions in the east; the problems of modern farming regulated by European Union norms and influenced by crop subsidies and international trade agreements, and modified by the access to western technology and other agricultural inputs; and the dominance of commerce by the west. Our interviews were complemented by observations and by a systematic reading of two journals dealing with agricultural issues in eastern Germany and the local newspaper. We were also able to avail ourselves of an ongoing study of agricultural enterprises in the region based on self-reporting of a sample of farms, historical analyses of GDR collectivization and the organization of agriculture, and the growing social scientific literature on agriculture after reunification.

In this book, we have decided to give as much of a voice to our informants as possible through the inclusion of long quotes from taped interviews. We thereby personalize the histories of firms and the meaning of economic and political transformation. This narrative strategy was dictated by our desire to capture the poignancy of the historical moment, which is quite unique with respect to the magnitude and rapidity of the change, and to record the multiplicity of ways in which these events and the more distant past were cognicized by our informants. The usual distance between observer and observed was in this case considerably narrowed through our common Germanic European background, level of education, and, in many instances, social class position. Also, our work with a relatively small number of farmers lends itself to a more individualized treatment of their voices. These voices do not, however, stand alone, for they are embedded in the understanding we gained by interviewing a wide array of other entrepreneurs.

Our aim, then, is to give the sense, to the extent this is possible, of a multiauthored text in which we see ourselves as facilitators in the telling of a common story by a number of individuals who have experienced and interpreted the events in parallel, disparate, and/or contradictory ways and who, in a very real way, are contributing to the construction of the future of agriculture in Germany. In order to minimize the privileging of the "author," we have decided to abstain from the usual practice of presenting quotes from informants in a different font or indented text. Each chapter attempts to forefront the voices of a particular category of farmers. Within each chapter we also highlight the differences among subcategories and individuals. This is not to say that we have not taken certain liberties with the interviews by translating, editing, and organizing them. We hope that our juxtapositions of quotes and our generalizations will be seen by our informants as emerging organically from what they have told us and, at the same time, that the rich material presented will enable the reader to develop alternative interpretations to our own.

The book is organized in two parts, the first dealing with the precommunist and communist antecedents of modern agriculture in eastern Germany and the second with the transformations after the watershed year of 1989 when Erich Honegger was removed from office and the Berlin Wall was demolished. Prefiguring the subsequent discussion, the two parts are preceded by a synopsis of the work histories bridging the two periods of three farmers representing the three major types of farmers presently heading farms in eastern Germany: the managers of cooperatives that emerged out of the GDR collectives, the eastern German independent farmers, and the western German farmers who moved to the east after 1989. In chapter 3, we discuss the situation in agriculture before communism and its transformation during the Soviet occupation between 1945 and 1949 and the subsequent changes in the German Democratic Republic. We argue that many of the traditional patterns as well as the innovations made during the communist period, even a few that were later largely superseded by other transformations, were incorporated into the forms that emerged after 1989. Chapter 4 analyzes the workings of agriculture in the GDR in the late 1970s and 1980s when the system had reached a considerable degree of stability. We argue that rather than constituting a unitary structure, the system entailed contradictory trends that enabled it to overcome some of its major deficiencies but also facilitated its subsequent demise. In chapters 5 to 8—which are preceded by an introduction to the transformations in agriculture that took place after 1989 with the fall of the communist regime and the subsequent reunification of the two Germanies—we take up the theme prefigured in chapter 2 and discuss the three types of farmers that constitute the main focus of the book.

Chapter 5 discusses the transformation of collectives into cooperatives with legal structures that correspond to western German models, a form chosen by most of the collectives that were not dissolved altogether, among various other corporate options because of its resemblance to collectives. We argue that beside the difficulties in compensating former members for the resources in the form of cash and animals and the practically rent-free use of their land, the need to dismiss a large part of the workforce, and the need to persuade current and former members to leave their land in the cooperative, the cooperatives have faced hostile government agencies that discriminate against them in various ways. In chapter 6 we discuss the changes and continuities in the functioning of these new cooperatives. Chapter 7 examines the eastern German farmers who have established private farms after 1989, farmers who themselves often had a background as cadres in GDR agriculture. Chapter 8 discusses the rationales for various kinds of western German farmers, both those with and those without family backgrounds in farming in the east, as well as for Dutch farmers to seek new opportunities in eastern Germany. In chapter 9, the discussion of farmers is complemented by a description of other actors associated with farming. They include tractor dealers and millers of both eastern and western German origin. The examples show that, although less pronounced, activities other than agriculture also entail some continuities with the past, but they also highlight the special position in agriculture both before and after the *Wende*. Finally, in chapter 10, we revisit and further develop some of the themes addressed in the book, particularly the specific views the different types of farmers have of one another, and speculate about the permanence of both the divisions and the congruencies among the various actors in eastern German agriculture.

CHAPTER TWO

Identities

To set the stage for the discussion of the various types of actors that have emerged in agriculture after 1989 and to give a preliminary characterization of the manner in which they identify themselves and are viewed by other types of farmers, we shall present three cases: that of Mr. Werner,[1] the former manager of a collective who now runs a cooperative; that of the Feldmanns, an eastern German family who reestablished a family farm; and that of Mr. Rascher, a western German farmer who also established an individual farm after the *Wende*. We shall present their cases from interviews taped in 1993–94, reserving the discussion of subsequent changes for later chapters.

We interviewed Mr. Werner in his office in an old, partially renovated farm building complex, the seat of a former demesne that now housed the cooperative's administration, a cow barn, and a new post-*Wende* architecture bureau. This was but the latest use for the building which, after the expropriation of large estates during Soviet occupation between 1945 and 1949 was used as a laundry, a distillery, and a polytechnical institute. Mr. Werner proudly pointed out that, while much of the complex was still in ruins, all the sections of the building owned by the cooperative were in good condition.

Unlike most cooperative managers, Mr. Werner did not have a family background in agriculture. He told us that he did not want to become a car mechanic like everyone else in his class, so, after finishing tenth grade, he decided that he had enough of school and started an apprenticeship in agriculture and then completed his training in trade school.

His first job was as an advisor in a tractor pool for agricultural cooperatives,[2] where he controlled milk quality. "I had received professional training so I was better qualified than those who had just recently come together to form cooperatives." After the tractor pools were dissolved, he first worked in a cooperative in the north for ten years and then decided to move closer to his mother. He took university courses for five years by correspondence. Although his idea had been to study agrarian law, he was persuaded by a classmate to join the collective, which became the cooperative where he was presently working. There he was first put in charge of pig raising, then of livestock production in general, and finally as overall manager, when his predecessor became ill and died soon thereafter at the age of 53. Becoming the manager took some doing because he had relatives in the west and was therefore suspect by definition. "I had to appear several times to be questioned whether I was up to the task, since I had relatives in the west. I was told that there were special rules governing such cases. They had their eyes on some other party militant. The board of trustees then came together in a separate session without me (I normally attended such sessions as production manager) and they decided that they wanted me and no one else. They did not want some stranger. They prevailed against the wishes of everyone else."

That this was no mean victory becomes apparent from another remark Mr. Werner made: "There were requests to break those kin ties with the west. I responded that my wife would regard that as a reason for divorce and that I did not intend to separate from my wife. Then, from time to time we noticed it in some indirect way and even in some not so indirect way. We were in X in June, 1989 for (my parents-in-law's) silver wedding anniversary. You see, I had already made a request one year earlier for my father-in-law's 70th birthday. The first time they told me to postpone the request for a year and I told myself, so what, I won't hand it in. But I did hand it in the next year for the silver wedding anniversary. Soon after we returned from the West, the collective was subjected to an unannounced control visit at 2 A.M. It could have been a coincidence, but maybe it was not. My wife told me, 'That is no coincidence, don't be so credulous.' I just answered, 'Oh well, it was our turn anyway, so what.'"

Mr. Werner remained in the position of manager ever since he was first appointed. Heading a livestock-raising collective was a challenge. "We were over 200 persons. We were separated (from crop raising). We were a pure livestock-raising unit with no land. We had to purchase feed from the crop-raising collective. You will recall that one of Erich's (Honegger) advisors had the 'glorious' idea in 1973 to separate field agriculture and livestock raising. Then there was this 'wise' pronouncement that crop raising had absolute priority. The rationale must certainly have been to reduce imports. Livestock production had to make do with whatever it could get from the crop units and as a result livestock raising was never effective. We had to maintain a certain number of animals regardless of performance. When the crop collective did not have enough to sell us, then each animal would simply receive less. Calf production was considered advantageous because of world demand for veal as well as sperm and embryo transfers." As a result, livestock-raising units were generally less prosperous than their crop counterparts and profit that could be distributed at the end of the year usually amounted to only a fraction of what could be distributed by the crop collectives. In fact, the latter often waived some of the payments for feed. But, in the case of Mr. Werner's collective, the economic power was reversed. As a result of inferior management, the crop collective with which it was associated had a bad name in the region, so bailouts went in the other direction.

Like most collective managers, Mr. Werner was a member of the Communist Party, which was important for his career. There was no particular pressure to belong to the Eastern German Communist Party (SED) (although very few managers did not belong to some party), and in Mr. Werner's community half of those with party affiliations and half of the town council belonged to the CDU (Christian Democratic Union). "One didn't have to belong. Many belonged out of conviction. Many, and those were the honest ones, had some political ideas of their own concerning certain issues. Then there were many who belonged for personal reasons, or to further their careers. It was easy to know who they were. Then there were many who belonged because it was easier to belong than not to belong. But those who really didn't want to join would quickly join the CDU or the *Bund der Deutschen Bauern* (BDB), whose programs, let's face it, were not really any different. The CDU celebrated Christmas and the comrades did not, but they also joined in the October 7 celebrations. I am sure that the relationships were somewhat different in the CDU than in the SED."

Belonging to the party entailed involvement in at least a minimum number of party functions. It also entailed the obligation to give truthful information about employees. This was often the case when the military came to recruit soldiers for the sensitive posts along the border with West Germany. But there was a degree of flexibility even there. "Truth is, after all, mutable. My truth did not have to be his truth. It was often very complicated to find a reasonable path in order to safeguard the members. One had to have some backbone and accept the fact that things could become disagreeable. Yes, one did have to have courage. In retrospect I think that we had too little. It could vary from too little to having a little too much. But as a rule we had too little." With respect to his involvement with the party, the fact that Mr. Werner had relatives in the west turned out to be a blessing in disguise, for he was not considered trustworthy enough to take on an active role in the party.

Mr. Werner was a member of the town council of the village of several thousand inhabitants, where he and four other SED members, five members of the CDU, and one representative without party affiliation came together to make decisions regarding matters concerning the community. "Such decisions were not made according to party politics, unless the comrades had a special agenda. In general, we dealt with local issues. Many of the comrades in the council had access to cash through the firms they represented. The kindergarten might need a new coat of paint and a new heating system. Who could have paid for that? The *Kreis* never had any money, and the communities as a result very little. So the collective would contribute 100,000 marks, and the galvanizing plant could contribute much less, but would take 20,000 out of the director's fund. We built a school and a gym. We built roads and a youth club which is now the hotel. All out of these funds. In ten years our collective spent 1.8 million for the community."

After the *Wende* the collective was split into three smaller cooperatives wherein livestock and crop production were rejoined. Individuals who owned land in the former collective[3] had the choice of initiating their own farms, renting the land to other individuals, or leasing it to the cooperative.

One of the three cooperatives failed and a new one was formed out of its ashes by eight of its former members while the other two, including the one that Mr. Werner heads, have been able to weather the transformation. He felt, though, that it in some ways it might have been better to dissolve the crop collective and start afresh, for now the new cooperative was saddled with the derelict unit's high debt, and while Mr. Werner was still hopeful that at least part of the old debts burdening cooperatives such as his would eventually be forgiven, this was yet to occur.

Mr. Werner and the other members of the executive board were elected by secret ballot by the full membership—an unusual procedure, for normally (i.e., according to the statutes of most cooperatives) the membership merely elects the board of supervisors, who then appoints the executive board including the manager for three-year periods. He explained, "From 1981 to 1990 I was manager and I put value on being elected by secret ballot, so that everybody could come to his or her own decision. I wanted to know if I would be elected or not. We didn't like the notion of appointment. We still do it that way every year. In September we will again have elections by secret ballot."

Interestingly, in spite of Mr. Werner's enthusiasm for democratic government of cooperatives, when we asked him directly, he was not sure that cooperatives would continue in their present form. "I believe," he said, "that we can maintain a firm with this type of organization. I don't believe that it (the cooperative) will continue to exist for many years. Probably another type will emerge, where a single person makes the decisions and can and must make them rapidly. Where I don't have to ask constantly whether people are in agreement, but where one person has the say. As far as the division of profit and the division of work is concerned, that will remain for some time, but I believe that who has the power to make decisions will have to change, because sometimes you have to make decisions rapidly. I can sometimes make a short-term decision and then ask people about it ex post facto, but if some new person would come in, he would soon try to organize things differently if he had such limited powers. So, I think that in this respect there might be changes, even though I don't find the present form bad at all. We are a good team here. We party well together. We can stay up all night when we have our annual celebration. [Others can't do that any longer]. Positions have hardened. People no longer see eye to eye in these times of change. The past certainly had many disadvantages but there was the very great advantage that everybody was held at about the same low level and so the feeling of solidarity was different. Now, I do have to say that after the *Wende* many mistakes were made in the different firms: accounts were made artificially low or were not presented to the members so people had the feeling that they were being cheated when it came to dividing [the assets]. At present there is a kind of war going on between the two associations of farmers [one representing mostly the cooperatives and the other, mostly the independent farmers]." Mr. Werner was to repeat these remarks regarding the viability of cooperatives in 1999, adding that if he could do everything over again he would either have started a farm of his own or a limited partnership (*Gemeinschaft mit beschränkter Haftung* [GmbH]).

Like many of our eastern informants, Mr. Werner also deplores the decrease in solidarity and neighborliness in daily life which, as we shall see later, resulted from the massive layoffs, increasing economic competition and differentiation, lack of voluntary associations to replace the more coercive forms of commensality during the GDR, and ostracism of former party operatives, among other factors. "This solidarity, the willingness to help one another, has decreased and I am sorry about that. I live in a double house in a row of houses. We have gardens in the back that face those of a second row. When we worked a bit in our garden on Saturday or Sunday mornings people would come out and jokes would go back and forth. Then someone would ask whether we really wanted to continue working or whether we wouldn't sit down for a glass of beer, tell a story, and play Skat. At 10:30 everybody would have finished working and we would sit down together. That doesn't happen any longer, or very rarely. Or in winter we would sit in the basement room that I had built for the purpose, talk and tell jokes, many political jokes, but that doesn't happen any longer either. A lot of people have made it in the new world and are convinced that they are doing things right. There is nothing wrong with that, but a lot of closeness has been lost. Before it didn't make a difference whether one was a member of the SED or any other party, as long as one was respected within the group, large or small. Of course, in the larger framework it did play a role, but not on the local level—I am referring to the village, of course, and not to the city. In that intimate sphere only character played a role. Today it makes a difference. All of a sudden people came to the fore who were not particularly noticeable before or some who also were movers and shakers in village affairs before. All of a sudden they belonged to the right party, quite by chance. Suddenly they behave differently. I don't know why."

The beginning was very hard for Mr. Werner. "During the first year I lost over 12 kilograms without having eaten one slice of bread less." Nobody knew what the next month would bring and in order to continue to be able to pay the workers who had to be given six to twelve months' notice, assets such as animals had to be sold off, sometimes at bargain prices. "We still paid every month. Many firms couldn't pay for as many as three months on end. Things were sometimes very difficult during socialism too, but this was unknown territory. It was much more difficult and complicated and there were a lot of people who enriched themselves at our expense."

However, at the time of the interview he no longer felt insecure. He felt that he had much more leeway to make decisions than before. Before the *Wende* the beginning of the harvest was dictated from above irrespective of whether local conditions warranted it or not. "[Today] I can decide whether to do something or not. When I feel that it is economically feasible then we discuss it briefly—one person alone doesn't always know everything—and then we leave it. If we think that planting sunflowers might be something suitable for us to do, then we first find out all we can about it. Before they would simply tell us that we should plant 15 hectares of this and 75 of that. And they would announce in the newspaper that such and such a collective had begun to thresh on July 1 and so it was time to begin too, otherwise the manager would have to appear in front of the authorities who would ask why he had not begun to thresh."

Mr. Werner's story presents in a nutshell the elements that shaped his present identity and that of other cooperative managers, most of whom managed the collectives out of which the cooperatives were formed. The managers' success was often predicated on a degree of political compliance, a compliance that positioned them favorably to continue managing the farming operation under the new rules, but also made them suspect in the eyes of the westerners as well as many easterners because of their previous collaboration with a repressive regime. As such, they were also under suspicion of treating their members, particularly those who wished to become independent, in authoritarian ways. In order to dispel this negative image, Mr. Werner and many of his fellow cooperative managers stressed the fact that managers of collectives rebelled against the often entirely arbitrary party dictates in myriad ways. Indeed, it is this independence and flexibility of mind that Mr. Werner and other cooperative managers regarded as an important qualification that have enabled them to run the farms under the new conditions. Mr. Werner, like other cooperative managers, stressed his previous role as a leader in his community, a role that he still upholds, but more in the narrower arena of the cooperative itself. These managers regarded as one of their principal goals the conservation of jobs in the face of massive rural unemployment occasioned by the need to compete in a capitalist system. Theirs was a quixotic position where they were obliged to reduce the workforce in order to survive, but tried to maintain a higher level of employment than their individual farmer counterparts by maintaining as many ancillary activities as they could and assisting some of their former members to become independent artisans.

Our second example is of a newly reestablished independent farm family, the Feldmanns. Theirs was a father-son partnership in which the wives were also involved. The family had farmed for many generations and had owned the family's present farm for two generations before they were forced to join a collective in 1960. At the time, animal raising was still in the hands of the family, but, in 1968, the farm became fully collectivized. The elder Mr. Feldmann first worked in the collective as a tractor driver and then studied a specialized aspect of farming and worked both in an institute and in the *landwirtschaftliche Produktionsgenossenschaft* (LPG—literally agricultural production cooperative but best translated as collective to denote the greater degree of collectivization that distinguishes it from western German style cooperatives). His son also became a specialist and taught agriculture at the university level while finishing his studies. The latter's wife had a diploma in agriculture as well. Training during the GDR was far more specialized than in the west. As the younger Mrs. Feldmann commented, "All-rounders (she used the English word) were not in demand. The idea was that each person should have knowledge of a specialized field, just like specialists in medicine." "The trick was," added the junior Mr. Feldmann, "to coordinate all these specialists, for that yet other persons were hired." During the years they were working for the collective, the family continued to raise a few pigs, which enabled them to maintain the house and the barns. The elder Mrs. Feldmann who first worked for a private gardener, later purchased eggs for resale to the state store in order to have more time to take care of her ailing parents-in-law.

The family decided to take their land (over 20 hectares) out of the collective as soon as the law made it possible and to lease between 200 and 300 hectares of land from individuals, the church, and the *Treuhand* (the agency in charge of privatization). This was difficult, since most of the members of the collective preferred to lease their land to the collective, so most of the land had to be leased from individuals who had never or no longer worked in agriculture. Once the leases were signed, the parcels had to be consolidated by engaging in exchanges with the collective. While in many instances such exchanges were made without changing the identity of the leaseholder, in this case the collective insisted on having the Feldmanns sign new leases. And they were forced to take over parcels with particularly complex property relations. Thus one small parcel that consisted of garden allotments had over a hundred owners who had to be found wherever they had moved, a process that took more than four years. "At the time, it was meant as a tactic to discourage us," concluded the junior Mrs. Feldmann. "They didn't think that we deserved it because my father-in-law had just come from the LPG. It was a competitive struggle."

The Feldmanns had little sympathy with the plight of cooperatives that were forced to continue to shoulder large debts incurred during communism, debts from which individual farmers and partnerships were exempt. They pointed out that the cooperatives did not have to start afresh, they were in possession of all the old assets and therefore did not have to take on all the *new* debts. Since the banks appeared to be particularly reluctant to lend money to agricultural enterprises, the money they felt that the cooperative still owed them was a particularly sore point. As the junior Mr. Feldmann explained, "After twenty-one years, I would at least like to get back the 2,700 marks per hectare of land that the farmers from Type I cooperatives (where only agricultural land was pooled) had to pay when they (were forced to) join Type III cooperatives (those where both land and herds were pooled). To date we have only received 1,700 per hectare. A thousand marks are still outstanding, (and that does not include the statutory 3% interest). And there are laws according to which the LPG has to pay 2 marks per *Bodenpunkt* (land point) per year, which would mean that for land rated at 100 points (i.e., the maximum), 200 marks per hectare would have to be paid for every year the LPG has used the land free of charge. Well, not a penny has been paid to this day." "Some cooperatives," added his wife, "decided that they did not want to engage in legal litigation. They said, 'We are decent and are going to support the farmers who want to start a farm of their own. We will pay those sums, have a pure conscience, and start the cooperative from the same level as they and then we can see who has the better enterprise.' But not in this *Kreis*. Here everything is brought to the courts. One has to fight (the German term she used literally means "to push forward with one's elbows"). I had to buy my first tractor and had to take a loan for a quarter of a million and then another quarter of a million because that was not enough. And my money is lying over there and the cooperative is working with my money with which I could have paid for the tractor. That's going on for four years. The psychological stress has become almost unbearable." The points of contention are not likely to be entirely resolved even in the longer run. For example, should the Feldmanns decide to raise cows, they would again be dependent on the cooperative to release part of their milk quota, for milk production is strictly regulated by the European Union, and another struggle could ensue.

But the Feldmanns did not regard the constraints arising out of the structures from the past as the only ones to affect them They were equally cognizant of the constraints imposed by the new systems they are entering. The Feldmanns' view of the European Union's agricultural policy is revealing. Although they complained about the top-down decision-making of the GDR, they were equally struck by the constraints imposed upon them by the present system. As the younger Mrs. Feldmann put it, "The extent to which the state dictates what can be done is enormous. . . . We can decide what to plant, but there are quotas. There are the sugar beet quotas, and in the dairy industry there is the milk quota. You are only allowed to plant this much of this variety and that much of that variety. As a result you are burdened with so many restrictions, that there is very little room to maneuver. In addition, there is the pricing policy. The precipitous fall of grain prices. One does, of course, try to do other things, in order to make things as profitable as possible. . . . And now the dictates regarding the amount of land that should be left fallow, all of which makes it difficult to maintain any kind of crop rotation."

The Feldmanns made no strict separation between state-driven and market-driven constraints, for the junior Mr. Feldmann interspersed his wife's comments with a discussion of constraints on production that were actually imposed directly by market demand, "From the vantage point of optimal cropping practices, one can't say that one is going to plant wheat and then barley, which would be the normal cropping sequence. If the price for sunflower seed was advantageous this year and it looks good for the harvest of '94 as well, then we have to plant sunflowers: 10 hectares of grain less and 10 hectares of sunflowers more. If in the following year the demand for rape seed is high, then we have to plant rape. Everything according to financial criteria. Cropping criteria become secondary." These limitations not withstanding, the Feldmanns found a lucrative niche producing potatoes for local consumption that they deliver directly to their customers.

The mountains of paperwork involved in finding plot owners and in obtaining bank credit, state enterprise subsidies, and European Union agricultural support engaged the younger Mrs. Feldmann some four to five hours a day. Her options were limited. She was able to take advantage of the GDR-mandated "baby year" with full pay when her youngest child was born. But the cooperative refused to rehire her since she was seen as the wife of a competitor, and the offers of employment she did receive were financially uninteresting. However, she felt fortunate that she could still work in her own profession, even if not, or at least not yet, in her special field of livestock raising. Her classmates had not been as fortunate. During a class reunion she found out that only 10% of her classmates were still engaged in agriculture. "They did not come from former farm families and did not have anything (i.e., land in the collective) and said to themselves, 'I don't have anything and so I will have to leave the cooperative anyway. They were also the last to have been hired, 87/88, and those who had come in last were the first to be let go." In 1999, her estimate of classmates who were engaged in agriculture was even lower. "Of the 200 students we had at the university, perhaps ten are engaged in agriculture. Insurance is very much in vogue. Well, if it were at least insurance in agriculture, that would still be alright. But many have become car salesmen and things like that." The elder Mrs. Feldmann helped with the children and also engaged in farm work, including work with the smaller tractors.

Used to frugality before the *Wende,* they first lived off the income from the younger Mrs. Feldmann's baby year payments and later from the small salary allotted to her for her work. Much of the rest of the farm's income went back into the farm. Indeed the younger Mrs. Feldmann considered frugality, and more generally, the capacity to make do with less than optimal situations as a virtue that distinguished her from westerners. She felt that the latter must always have their wishes fulfilled even when their economic situation does not warrant it. She was amazed how spoiled western students are compared to the Spartan conditions she had to endure during her own studies. She was also appalled how social differentiation is increasing under the new capitalist conditions and how grade school kids who were just a bit slower than others were already left behind in the competitive struggle, paving the way for increasing class differentiation.

The Feldmanns, then, shared a cadre background with the cooperative managers as did most of our other informants from eastern Germany who established private farms or father-son partnerships. They brought to bear a background of specialization in specific aspects of agriculture but also experience engaging in more varied aspects of individual farming. The Feldmanns, like many of their peers we interviewed, had once been private farmers before they were forced to join collectives or had joined them voluntarily. In addition, most of them had engaged in small-scale individual livestock raising and farming on small plots allotted to them by the LPG. Their identity was also formed in opposition to the LPGs, which often laid obstacles in their path toward setting up their own farms. In other instances, they also defined themselves in opposition to the western farmers who tried to hire them as managers of farming operations that they were attempting to institute in the east. They claimed that one of the reasons they had established their own farms was to demonstrate the capabilities of eastern farmers. Their identities were also shaped by their insertion into a new system, one that provides diffuse incentives that are combined with a highly centralized system of subsidies and limitations imposed by the Common Agricultural Policy of the European Union and by the German state and the federal state *Land*. Similar to the cooperative managers, they were inserted in a developing class system that was increasingly isolating them from their former friends and colleagues. Finally, with the loss of employment for women outside the home, the family became even more important than it was before the *Wende*, when it already played an important role as a refuge from an intrusive regime and provided a basis for the ancillary household-based farming economy.

The final case history is of a western German farmer who came to Sachsen-Anhalt soon after the *Wende* and established a farm of over 300 hectares. Several considerations influenced Mr. Rascher's decision to establish himself in the east. First, although he was likely to inherit the 100 hectare farm in the west, his father was still relatively young and did not need his full-time help on the farm. So Mr. Rascher decided to become a farm administrator for someone else.[4] After a short stint in the United States, he could still not find satisfying work in western Germany. Second, the collapse of communism had the immediate effect of opening the borders to competition from the west, which very rapidly led to the downfall of such heavily subsidized areas of GDR agriculture as sheep raising, leading in turn to massive sell-offs. The Raschers' sheep-raising operation declined instantly, which constituted a further disincentive for the younger Rascher's participation in the family farm. At this time, he was approached by a distant eastern German relative who urged him to take over the farming of his land. The families had been in contact throughout the GDR period through frequent correspondence and occasional visits. Since Mr. Rascher had just earned a degree as master farmer after finishing a two-year training program at a trade school, he took his relative up on the offer.

Mr. Rascher's initial strategy was to hedge his bets. He rented additional land in the east but continued commuting between his father's farm and his new operation. Even in 1994, he continued to live in very modest quarters (he called it "student housing") in the east. Personal considerations also contributed to his reluctance to make an immediate move to the east. His girlfriend, whom he later married, was still completing her studies and student teaching in 1993–94. While he initially still helped his father during peak periods, Mr. Rascher had to spend more and more time in the east both with the mounting paperwork and agricultural tasks. After a year, he and his father decided that the farm in the west would eventually have to be leased to others, while Mr. Rascher would move permanently to the east and build up a farming operation entirely based on crop farming.

At first, obtaining land in the east was quite easy. "At the very beginning people were interested. They were open to new possibilities. After all, the lessors who were no longer members of the collectives received little or nothing for their land. As a result, I obtained a lot of church land. The church was amazed that one could obtain a lot of money per hectare." Some of the collectives saw the handwriting on the wall too late, "while others adapted quickly, kept much of their land, and are in a good position today." Soon thereafter he encountered more local resistance. "It is true, that I was taking away land from the LPG by deciding to settle here. I had to act very carefully not to give the impression that I was taking away work from the people; for at least 60% of the people in the village must have been working in agriculture. (In the GDR) employment in the collectives was subsidized. As a result there was no unemployment. One can quarrel about whether that was good or not. . . . In the west one calculated an average of one worker per 120 hectares. Here it was five or six times more workers. The LPGs proclaimed that the western Germans were coming to take away the land which would mean that the LPGs would have to let people go. Bad politics that! As a result it was difficult to become established here in the village." With one former collective he had a good relationship and was able to engage in land exchanges with relative ease. In contrast, another cooperative manager made life as difficult for him as possible. Mr. Rascher felt that this manager had been trained under socialism and saw the private farmers as an example of the social market economy and was therefore simply unwilling to cooperate with western farmers.

Mr. Rascher enjoyed certain advantages over his eastern colleagues. Unlike some of his colleagues from the west he did not have much start-up capital of his own, since his father still had his own operation, but he could bring sufficient used equipment from the family farm to tide him over the first years. He was able to invest an additional quarter of a million marks in the form of bank credit and leasing arrangements. At the time of the interview, he rented a farm building from the *Treuhand* and had purchased a second one. In addition, he bought an old house in an attractive location which he planned to renovate.

In the three years since he started farming in the east, Mr. Rascher became sensitized to the differences in farming conditions and in the politics of farming between the two regions. The severe drought of 1992 taught him that he should have started the farm on a larger scale, taking advantage of the greater ease of obtaining land early in the game to provide a cushion for such lean years. He felt that western farmers like himself who had come to the east had taken insufficient account of such factors as the Harz Mountains, which cast their rain shadow over the region, and the relative ease with which the fertile loess soils could be plowed. He was also thinking about the future when agricultural subsidies might be curtailed. "I believe that in order to cushion the effect of lower prices—the price/cost scissors will separate more and more—one will need more agricultural land. That means we must enlarge. The motto is: grow or move out."

There were also restrictions of an agro-political nature. His father had the good fortune of having moved heavily into sugar beet farming before the quotas were set for sugar beet production, a crop that is particularly lucrative at present. In contrast, the quota was low in the region where his new farm is located. He realized that such quotas were artificial devices that were dependent on the clout of the farmers in the sugar factory. A factory co-owned by the farmers tended to be more responsive to the farmers' needs than an enterprise with less than 50% participation, such as the brand new sugar factory near Bernburg, which monopolized production of a large region. This fact had meant that while sugar quotas could be exchanged for other production quotas among producers in many regions in the west, this was not possible in the Bernburg region.

However, the most difficult part of establishing a farm in the east was the impossibility of obtaining a telephone line and the difficulties of dealing with officials who were unfamiliar with the tasks imposed on them by the transformation. "It was quite an adventure as a private farmer, especially for people coming from the west who were unfamiliar with conditions around here." There was also the fact that, unlike in the west where small and medium farmers often cooperate by sharing equipment, the newness of private farming and the large scale of operations has meant that everyone was forced to rely mostly on his or her own resources. Mr. Rascher attributed his success at adapting to the fact that he was willing to make do with rented lodgings and with an older car prone to flat tires because of the bad road conditions. He found it important that he could get along easily with others and did not put on airs. "Others perhaps still had an old farmstead where they could return. But people who were not so well connected, it was very, very difficult. Life here is not for everyone. The differences with the old *Bundesländer* are great." He believed that because both he and his wife were adaptable, have simple tastes, and enjoy such pleasures as an old house built out of stone in the ancient local tradition and the rural landscape, they would both be able to make a life for themselves. At present, although he did meet with eastern farmers in a farmers' association, most of his social contacts continued to be with westerners. He hoped that that would change when his wife joined him and found a job in her profession. He was encouraged in this belief by the warm and helpful manner with which he was received locally, in spite of the negative experiences the easterners had had in recent years with western swindlers and profiteers.

For Mr. Rascher and the other western farmers we interviewed in the east, eastern Germany represented a new frontier, a place where they could have exotic experiences and where they could realize dreams that had remained unfulfilled in a society where there was little room to engage in grand new ventures. A colleague of Mr. Rascher, a man of *Junker* (noble) background, said to his west German manager in makeshift quarters in a crumbling mansion: "Das erspart Ihnen ein Jahr Afrika oder ein Jahr Südamerika. Das Abenteuer haben Sie dann hundert Kilometer von Ihrem eigenen Hof entfernt. So nah. So schön wird's nie wieder." ("This can take the place of a year in Africa or South America. You can enjoy an adventure [only] 100 kilometers away from home.

That close. It will never be so good again."). Another western farmer, also of *Junker* background, used almost the same words: "For our generation this is a fantastic opportunity. We have no need to emigrate, to be pioneers. We can do it in our own country, using our own language and on familiar grounds. That is really a wonderful situation." For Mr. Rascher himself, however, the move was more than an adventure. He could establish his own farm rather than having to wait until his father's retirement. The westerners interviewed also all came from relatively privileged backgrounds. Mr. Rascher was typical in this respect. His father owned 67 hectares of land and rented an additional 25 in an area of high fertility. The family's bourgeois credentials are underlined by the fact that his brother was a physician. Family connections in the west also facilitated access to credit. In sum, these informants were unlikely to fail and if they did they had plenty of backup options. Finally, the westerners' identities were shaped by struggles of their own to find acceptance in the east, overcome the loneliness of separation from spouses, family, and friends, and face the often hostile reaction of the cooperatives with whom they competed for land. But the westerners were perhaps more heterogeneous than the other two categories of farmers explored in this book. Some, like Mr. Rascher, came from long-standing small- to medium-scale farm families. But others, like the ones quoted above, were *Junkers*, noblemen, who were regarded with some suspicion by their fellow westerners. Some of the latter cultivated the land themselves. Others sent farm managers and rarely appeared on the scene. Finally, there were large-scale corporations that purchased or rented land in the east.

While the meanings of the cases above will be elucidated when they are contextualized in later chapters, it will already have become apparent to the reader that the actors are all presently involved in major struggles with members of the other categories. As we shall see in later chapters, these struggles in many ways define their identities and in some instances their very existence. The often inconsistent definitions of identity—Mr. Werner's as independent and flexible, the Feldmanns' as oppositional, and Mr. Rascher's as both adventurous and cautious—were developed by old and new economic and political realities. Even from a cursory comparison of the three cases, it is clear that these actors have been shaped by very divergent histories and that when they relate to their historical past they draw their inspiration from different time periods: the heyday of the LPG in the 1970s and 1980s in the case of the cooperative manager, a broader historical range from the family's history as independent farmers before collectivization to the enculturation as specialists during the collectivist period for the independent eastern farmer, and an attempt at tying into the family's precommunist and even prewar past in the case of the western farmer. It behooves us then, as a first step in our analysis, to briefly outline the region's history in general and the history of agricultural collectives in particular, and then to examine the opportunities and constraints faced by the managers of collectives (a category to which most of our independent eastern German farmers also belonged) in agricultural collectives during the GDR.

PART I

The Communist Period

CHAPTER THREE

The History of Collectivization in Agriculture

As we have seen in the introduction, the history of land tenure and holding size could be characterized as layered. Even after major transformations, later periods often repeat earlier patterns. Thus post-*Wende* capitalist agriculture borrowed from both prewar and GDR patterns. The same layering is apparent during earlier transformations. Thus, after a period of fragmented landholdings resulting from the agrarian reform of 1945, instituted by the Soviet occupation force, there was a return to large-scale landholding patterns. The stages of this transformation are important in understanding the post-*Wende* changes because some of the transitional forms as well as some of the experimental forms that never became dominant in the GDR have been incorporated in the emerging forms. The history of agricultural cooperatives is also one of resistance and of subsumption of recalcitrant forms into the logic of socialism, a subsumption which, although largely successful, was never complete. We will argue that it is precisely in this space between ideal forms as imagined by the state and the flexible ones that the members and managers would have preferred, that we find both the plethora of solutions that enabled GDR agriculture to survive and even to prosper relative to other sectors, and the seeds for the evolution of new forms that emerged after the *Wende*.

Sachsen-Anhalt, the site of our study, and, more specifically, the region between Magdeburg and Halle, where the *Kreis* of Bernburg is located, was considered an important agricultural and industrial region of the German *Reich* (state) since its foundation in 1871. Indeed, agriculture was among the most developed in the country (Mende 1994:41) and held this position during the entire pre–World War II period (see Kramer et al. 1957:12–15). In this chapter, we introduce the convoluted transformation of agriculture in Sachsen-Anhalt, a region characterized by long-term family traditions of large-scale mechanized privately owned or leased estates that were expropriated after World War II and distributed primarily to landless farmers and refugees, then reorganized into cooperatives and further collectivized along a number of models that sought to both concentrate and separate functions. In the worlds of farm workers, the different state policies and experiments brought trauma and challenge, compliance and circumvention.

Historically, agriculture in Sachsen-Anhalt was practiced in holdings of varied size, but with a much larger share of the land held in large units over 200 hectares than in other parts of the *Reich* (see Table 3.1). The table shows the distribution of cultivable land between different size categories in 1925 (Berndt 1994:109).

By 1939 this distribution of holdings had barely changed and was comparable to the holding size under communism in eastern Germany.[1] Compared to the structure of landholdings in western Germany, far more land (by a factor of 7) was held in holdings over 100 hectares and less than half as much in the form of holdings under 5 hectares (Kramer et al. 1957:10–11). However, as Table 3.1 shows, very small operations accounted for a somewhat higher percentage of the land than the national average. These farms belonged to part-time farmers who also worked in mining and industry (Berndt 1994:107–109).

TABLE 3.1
Size of Farms in Sachsen-Anhalt and in the *Reich* in 1925

	Administrative Unit		
	Magdeburg	*Merseburg*	*Reich*
under 2 hectares	7.10%	7.30%	6.25%
2–5 hectares	5.30%	6.80%	11.40%
5–20 hectares	29.50%	29.00%	35.80%
20–50 hectares	23.80%	20.90%	19.80%
50–100 hectares	10.40%	9.40%	6.60%
100–200 hectares	5.80%	8.20%	4.90%
over 200 hectares	21.70%	18.40%	15.30%

Source: From Berndt 1994:109.

Some of the farms' holdings were cultivated by those who owned them outright or who hired others. Other land was held under long-term leases. Since under German law leaseholders received considerable protection and leases could not be broken unilaterally even when ownership changed, this form of access to land has been and continues to be popular in the region for farms of all sizes. For example, in the 1920s, a sugar factory in Klein Wanzleben farmed 2,900 hectares of land, of which 2,000 were leased (Mende 1994:51). Before the Soviet Union occupied eastern Germany, large estates, which occupied 30% of the arable land in the highly fertile Magdeburger Börde and the Altmark, included both those of the *Junkers*, or nobility, and of bourgeois families that predominated in Sachsen and state demesnes (*Staatsdomänen)*, which predominated in Anhalt (Berndt 1994:107). The demesnes were estates that had once belonged to the Duchy of Anhalt, and with the revolution of 1917 became the property of the *Land* and continued to be leased to farm managers on a long-term basis. Indeed, such leases could even be passed on to a farm manager's son. Improvements made by the leaseholder were protected, so *Domänen* renters treated such estates, including the buildings on them, almost as if they had owned them outright. The nobility cultivated half of the land itself and leased out the remainder (Hübner 1994:20). Land belonging to the church was also cultivated through long-term leases.

Agriculture in the region of the province of Sachsen was highly mechanized even in the late nineteenth century. By the turn of the century, the province boasted one-third of all the steam plows in the *Reich* (Hübner 1994:20). By the 1920s, agriculture was often integrated with industry. In 1928, sugar production in the region accounted for one-fourth of the sugar produced in the German *Reich* and more than one-fourth of wheat and barley (Berndt 1994:107).

As in other regions of Germany, farmers often passed the land on from generation to generation. And, as we shall see later, those eastern German farmers who established independent farms also often came from families with farming traditions similar to farmers in the western regions.

This system of large-scale farming underwent major changes after World War II, when larger estates were expropriated during the Soviet-instituted land reform of 1945 (known as the *Bodenreform*) and parceled out among refugees from former German territories annexed by neighboring Soviet bloc states and other land-poor farmers.[2] According to a former manager of an agricultural collective, "First, all farms with more than 100 hectares of land were expropriated across the board. In addition all those (some 4,000, according to Kramer et al. 1957:17) who were regarded as Nazi criminals and those who had committed some crime or were considered to have committed crimes [during the war also lost their land]." The expropriated land was apportioned in parcels whose size depended on availability and demand in a particular locality.

The same consultant quoted earlier explained the process as follows: "If there were, say, 100 hectares of land available and there were 20 petitioners, then each only received 5 hectares, but if there were only 15, then they received somewhat over 6 hectares. But it also happened that not all wanted the maximum of 10 hectares. Each individual could decide, for property also comes with obligations and it was not easy to farm the land at the time. There were no machines and no draft animals and [the petitioners] had no means at all. With the land each new farmer received a piece of the large barns belonging to the large [expropriated] estates.[3] They made partitions or divided [the buildings] in some other way. It was not easy."

As Table 3.2 shows, preference was given to landless farmers and agricultural workers and to resettled persons. The largest share of this land went to landless farmers and agricultural workers, followed by almost 90,000 refugees from former German territories (Kramer et al. 1957:19). "While they did receive something," the collective manager continued, "the farms they received were not really functional units, but parts of a larger operation, and now they had to create complete small farms out of these parts, which most did not succeed in accomplishing. As a result, many of these operations never became profitable: that is, as soon as they received the farms, the farms began to deteriorate." Indeed, by 1953, one-third of the new farmers had already given up their farms (Tümmler et al. 1969:30). That land went back into the state-owned

TABLE 3.2
Allocation of Land in the Land Reform of 1945

Recipients	*Total Hectares*	*Average Amount Received*	
119,530 landless farmers and agricultural workers	924,365	7.7	has
80,404 land-poor farmers	270,949	3.4	has
89,529 resettled persons	754,976	8.4	has
45,403 small tenant farmers	43,969	1.1	has
169,427 workers and artisans	111,203	0.7	has
62,140 established farmers (forest)	62,140	1.6	has

Source: From Kramer et al. 1957:19.

pool of land and was either cultivated in the form of state farms (VeB or *Volkseigene Betriebe*) or was given to cooperatives to use; for the beneficiaries only had usufruct rights over the land they had received rather than owning them outright (see Hartmann 1995:10–11).

The decision not to collectivize immediately followed the Russian model. Immediate collectivization may also not have been politically feasible at the time. One view is that the creation of very small farms was a deliberate strategy to force eventual collectivization without incurring the cost of antagonizing the rural population. One specialist in corporate structure who became a tax consultant after reunification commented, "I don't like to say it but one must observe history truthfully. The assumption behind this Lenin-inspired plan for cooperatives was that small farmers would join cooperatives." In addition, giving land to the refugees was a means of integrating them into the economy and preventing possible future unrest. Giving land to individuals also reduced unemployment in industry resulting from the destruction during World War II. At any rate, in 1945, the Soviet occupation authorities merely ordered the reconstitution of previously existing agricultural purchasing cooperatives, the *Raiffeisen-Genossenschaften*, and allowed the formation of other forms of voluntary organizations, all of which were dissolved again later (Kramer et al. 1957:21). They were eventually replaced by the much more centrally controlled *Vereinigung der gegenseitigen Bauernhilfe* (VdgB), the "Farmers' Association for Mutual Assistance," instituted in 1946 (Horz 1961:78–81; Ludz 1979:449–450), whose initial task had been to organize and manage farm machine stations where individual farmers, and later collectives, could borrow old equipment that had been expropriated during the land reform, to which some new equipment was added. The machine stations (*Maschinen-Ausleih-Stationen* [MAS] were later renamed *Maschinen-Traktoren-Stationen* [MTS] after they were concentrated in larger units and effectively had come under public control) also followed a Soviet model. Later, these stations also gained cultural functions and became tools for advancing the party agenda in rural areas, including the occupational and political education of members of rural collectives and the elimination of larger-scale farmers from political office (Horz 1961:76; Tümmler et al. 1969:41–45). Since private farms did not have direct access to farming equipment, even the larger farms became dependent on the MAS/MTS, or had to make do with aging equipment.

In 1952, the government instituted a new form of cooperative which, unlike the traditional *Raiffeisengenossenschaften*, was geared toward production rather than merely commercialization and the procurement of inputs. Depending on the type of cooperative, its members pooled only the agricultural land, keeping their animals and machinery

(*Landwirtschaftliche Produktionsgenossenschaft [LPG] Typ I*), also pooled machinery (*Typ II*), or held all means of production in common (*Typ III*). In all three types, members retained ownership of the land they brought into the cooperative. Indeed, unlike such communist countries as Bulgaria (Creed 1998) and Rumania (Kideckel 1993), land not expropriated during the land reform of 1945, or forfeited by beneficiaries of that reform who ceased to cultivate it, continued to be individually owned. The structure, rights, and obligations of these units were highly codified and subject to state control (Tümmler et al. 1969:57). In these cooperatives, the distribution of income remained partly determined by the amount of land brought in by each member. In Type I LPGs, as much as 40% of the income could be distributed according to the amount of land brought in. The corresponding figure for Type II collectives was 30% and for Type III cooperatives, 20% (Tümmler et al. 1969:62–63). Small-scale private production was allowed in all cooperative types, with each member household receiving half a hectare of land for personal needs that could be farmed individually or jointly. Each member could raise up to two cows and two sows with offspring, five sheep, and an unlimited number of goats, chickens, rabbits, and other small animals (Tümmler et al. 1969:82). While stressing the voluntary nature of these arrangements, the government favored the cooperatives from the start, giving them preference in terms of access to the machine pools, free advice from agricultural extension services, a two-year tax holiday, access to credit, deliveries of seed and consumer goods at low prices, preference in obtaining construction materials, and additional subsidies for those LPGs that took over land that had reverted back to the state as a result of the flight of the owners to the West or land that had been relinquished by agrarian reform beneficiaries (Kramer et al. 1957:50, Tümmler et al. 1969:90–91). Individual cooperative members received additional benefits.

Pricing policies for agricultural products and inputs had already become a means for directing change in agrarian structure. The farms with over 20 hectares of land that remained after the agrarian reform of 1945 were, at first, only slightly discriminated against in the compulsory sales at low fixed prices that the occupation government instituted from its inception. However, by the early 1950s, these larger private farms were placed in an increasingly disadvantageous position, having to deliver a higher percentage of their output at low rates, leaving them with less to sell at higher rates or to dispose freely in the specially constituted farmers' markets. This fact contributed to the flight of these larger farmers to western Germany (Berthold 1972:62).

Initially, larger farms were not allowed to belong to cooperatives, but this rule was changed in 1955. While the number of larger private farms decreased, especially during a campaign to discourage larger-scale private farming in the early 1950s, cooperatives progressively increased in number from 1,906 in 1952 to 6,047 in 1955, and 10,465 in 1959 (Tümmler et al. 1969:87). As one of our consultants explained, while most of the farmers who joined cooperatives were small and often the less successful farmers, some of the larger, more successful ones joined because they thought they might be able to manage the operation and benefit from the larger size of the unit. On the other hand, no cooperative was ever allowed to fail. One former LPG manager explained it this way: "At least in the first year, if someone worked well, he was better off working alone than joining an LPG. Most LPGs did not earn well. Of course, there were always exceptions, there was a lot of variation. There were very rich and very poor LPGs. During socialism, no one was swept away; the state compensated [their failure] in one way or another. First they would receive credit and all sorts of things. Then, at one point or another, the debts were pardoned and that was that. But as the LPGs developed the young people became used to them. The LPG had many advantages. Unlike private farmers, those belonging to LPGs had regular working hours. Each farmer had vacations."

Practically all the remaining farmers were strongly "encouraged" to join cooperatives during an all-out political campaign in 1960. The same consultant remembers this campaign with some embarrassment, feeling that it tainted a process which, in his eyes, otherwise enjoyed legitimacy. He described the development of LPGs as follows: "The cooperatives were conceded a lower delivery quota, so that, if they worked exactly as well as the farmers, that is, if they harvested the same amount, they had a higher proportion of freely disposable production (*freie Spitzen*). This was a material incentive for the development of cooperatives. So there were more and more of them and in 1960—we call it the Socialist Spring—there was a great political offensive, during which all sorts of people traveled around to the villages to try to generate enthusiasm and win over every last farmer for 'the cooperative path,' as they called it. They did this in all sorts of ways. First through friendly conversations. Of course, there must have been cases—it is only human that this should happen—where there were people who would say, 'If you don't cooperate you will land in Siberia,' and things like that. Such things must have happened, but there never was a legal basis for such statements. The proof that no one was forced to join is that a few farmers never did join a cooperative. Maybe it was only one in 500, but there were individual farmers until the very end in the GDR. People did try to blackmail the farmers a bit and to drive them into a corner. They also used enticements such as promising a car during a time when cars were so scarce that one had to wait ten years to obtain one. A farmer might be told, 'If you sign here and join the LPG, you will receive a car next year.' However, those who absolutely refused, did get away with it. But there were only very few."

Later in the interview he did concede that it was very difficult for his parents to join the LPG in 1960. "My father was subsequently seriously ill for an entire year before he got over the shock," he conceded. "But ten to fifteen years later we were all convinced that it had been the right path. It was a very different kind of life than working 365 days a year. To be sure, in winter we didn't always have to work eight hours a day either. But we always had to get up at five in the morning, at the latest. We had to milk the cows, and feed them even in winter. My parents never went on a vacation for more than one or two days, and that only when I was older and could take care of the farm myself if necessary, or much earlier, when my grandparents could take over. So, after ten to fifteen years, when people working in the LPG were beginning to make some more money, even my parents said that it was good that things had turned out they way they did. They said, 'We work eight hours, have enough to feed and clothe ourselves, and can afford a normal amount of comfort.' It wasn't possible to take long trips. There wasn't any demand for that either, at least not on the part of my parents. They had never known anything else. We had become used to [the new situation]." The West German literature on the subject paints the events of 1960 in darker colors. Psychological methods included the use of such slogans as: "Those who are for the LPG are for peace! Those who are against the LPG are for war." Indirect pressure included trumped-up accusations of having committed a punishable offense and a promise of pardon if the accused joined an LPG (Tümmler et al. 1969:93–94).

Many farmers who were forced to collectivize in 1960 formed Type I cooperatives, which permitted them to continue raising all animals privately. One farmer explained: "My father returned from war imprisonment in 1948 and took over his father's farm. He tried to remain independent as long as it was permitted by the state, that is, until 1960. In that year, the so-called Socialist Spring was announced and every other day someone from the *Kreis* came to pressure the remaining farmers in the village, there were three farmers in addition to my father, to form a Type I cooperative at least. My father always had about sixteen cows and a hundred pigs [which he could keep]." Indeed, most farmers appear to have either formed Type I or Type II cooperatives, or joined existing cooperatives. Between June 1959 and May 1960, the average size increased by 38.6% to 154 hectares in the case of Type I and Type II cooperatives, and by 68.2% to 534 hectares in the case of Type III cooperatives. Interestingly, few if any new Type

III cooperatives were formed in the period of forced collectivization. While the number of cooperatives of Type I and Type II increased from 3,018 to 13,033, that of the more fully collectivized Type III cooperatives actually *decreased* from 6,548 to 6,323 during that period (see Tümmler et al. 1969:87, 94).

Later, in the 1960s, Type I and Type II cooperatives were pressured to transform into the more highly collectivized Type III form. An independent eastern German farmer whose father had to join a Type III collective in 1970 described the pressure techniques used to force the switch from a lower to a "more advanced" form of collectivization as follows: "It started with the special feed for the milch cows. We would no longer have received any. As for fertilizer, allocations to us had always been limited. The same thing was true for technology. We had money on an account. It was called an indivisible account, which every Type I cooperative had to open. In 1970, there were still 60,000 marks on that account, but we couldn't spend it because we couldn't obtain technology. The allocation was made from the *Bezirk* to the *Kreis*, which gave preference to the [Type III] LPGs." Between 1961 and 1966, the number of Type I/II cooperatives decreased by one-third to 8,157, but presumably mainly through consolidation into larger cooperatives of the same type, while the number and average size of Type III cooperatives had barely changed (Tümmler et al. 1969:156). However, by September 1970, the number of Type I/II cooperatives had dropped to 3,485. Five years later, only 306 remained. Although their average size had increased to 302.2 hectares, they represented only 1.9% of the farmland (Ludz 1979:654). One *Wiedereinrichter* described the final phase of collectivization as follows, "We still experienced the last throes of private farming. I saw how the animals were driven from our barn. After it happened, my grandfather was sick for weeks. I can still picture it vividly. My grandfather's herd already reached an average of 7,000 liters [per year]. He had registered cattle and pig herds. The cows all had horns. When they became part of the cooperative they had to be without horns. So, during a brutal operation, people sawed their horns off." A former LPG manager had a more benign interpretation of the transformation of Type I/II into Type III cooperatives, "The Type I cooperatives slowly also joined Type III cooperatives. This trend was related to the aging of the members. Taking care of the animals at home became too much for them. Their children had found work in industry and so they had no other option but to join Type III cooperatives. The last Type I cooperative joined us in 1975."

The joining of the cooperatives into larger units combined with the standardization of the form of cooperatives initiated the process away from cooperatives and toward collectives. That process entailed the reduction of differences both among individual farmers and among farming units. The trend was toward redistribution of resources by means of more centralized allocation and the progressive reduction of decision-making power at the local level.

While the combination of LPGs into larger units that characterized the 1960s could have had the effect of creating economies of scale—and indeed, unlike their western German counterparts, who generally believe that units of much over 1,000 hectares become unwieldy, eastern German farm managers feel comfortable with 3,000-hectare and some even with 6,000-hectare farms. The practice also was used as a leveling device between more successful and less successful cooperatives. One former LPG manager explained: "The differentiation among the individual farms would have been much more pronounced [without consolidation]. The SED possibly decided to combine smaller units because, in spite of a planned economy and all the supports that weaker units received, there was a clear trend toward differentiation. There were very good farms that became better and better, in fact, so good that the people in the *Kreis* councils and in the ministry in Berlin were already becoming uneasy, and there were others that functioned badly and became worse. What do you do in a situation like that? You might say that you could throw this one together with that one, because there is someone in that one who is capable. It also cost the state money, to constantly provide support. This practice did not work out well. On rare occasions, I did get to know large farms that were outstanding; where, as a result of the combination of a good and a bad farm the bad farm soon reached the level of the good farm because the manager of the bad farm was subordinated to the manager of the good farm who knew the business and organized everything. But I saw many others where that result was not achieved at all."

The final major push toward concentration of land into large units (with an average range of 4,000 to 6,000 hectares) came in the 1970s, first with various forms of cooperation among LPGs that made arrangements to engage in complementary specialization and exchange of inputs, or pool plant production, and climaxed in 1978 with the systematic separation of plant and animal production into separate units (Ludz 1979:656–662). The animal production units were, in turn, somewhat specialized as well. Some raised only milch cows and sheep, others, pigs and beef cattle, and so forth. The consolidation of animal production units into larger specialized LPGs enabled the state to get rid of its black sheep in much the same way as the earlier consolidations. According to one former LPG manager, "J. was a very rich cooperative. It had some 3 million marks in a bank account. That was a considerable sum in GDR times. It didn't bring in a lot of interest, but it brought in some and thereby the amount grew further, because it wasn't used. The neighboring LPG had debts in the amount of 2.5 million. Then someone had the glorious idea of separating plant and animal production. At the same time, several animal production units were combined. Much against the will of J., the animal production cooperative Q. was added to it. It was a beautiful arrangement. The bank got rid of a 2.5 million marks debt and the animal production cooperative J. was left with 500,000. The mathematics were perfect. The bank had no further worries."

As we shall discuss more fully in the following chapter, the separation of plant and animal production was regarded as a major mistake by all of our consultants, regardless of whether or not they identified with collectivized farming. At the time, unlike earlier periods, plant production was generally favored over animal production, resulting in major income disparities among individuals who had once belonged to the same LPG. To quote one former LPG manager, "[The allocation of individuals to plant or animal production] was arbitrary. They looked for people who knew something about plant production. . . . Consolidating plant production was something new. . . . No one objected [to the fact that those who worked in animal production no longer had any control over their land]. Land had no value, it had no price, so it could not become a point of contention. Now, things are totally different.

"Things were really bad. It also produced a lot of discord within families. A husband might be a tractor driver and be assigned to plant production while his wife remained in animal production. At the end of the year, when profits were distributed, the husband, who actually had a less strenuous job, received a bonus of 4,000 to 5,000 marks, while his wife, who had worked for 300 days in animal production, received 800 marks if she was lucky. No one was asked where he or she wanted to work. People were placed according to the tasks they had done before. Those who had worked in a workshop were given a choice of going here or there and so they would land in either plant or animal production. [I didn't know which would be better], otherwise I would have joined plant production."

The state experimented with even further consolidation. In each of the fifteen *Bezirke*, five or so LPGs in one of its *Kreise* were joined into a gigantic unit called *Agrar-Industrie-Vereinigung* (agro-industrial complex) that also included related industries such as herbicide or barn construction factories. Unlike in Bulgaria, where farms were temporarily grouped into vast "agro-industrial complexes" (Creed 1998:74; Wenkart

Smollett 1984), that stage was never generalized, but the idea of joining different activities was applied more broadly by encouraging persons other than farmers to join the cooperative in order to diversify its activities. Thus, some artisans who engaged in one of those rare activities that did not become entirely public, also were, on occasion, pressured to join LPGs. The following example of a butcher who was cajoled into running an LPG-run butcher shop shows both the pressure tactics used to manipulate individuals and the manner in which LPGs became involved in ancillary activities, a trend that characterized the 1970s and early 1980s.

Mr. Herder, who presently runs a highly successful meat-processing operation, had lost the right to work as an independent butcher in 1975 as the result of a trumped-up charge regarding his subsidiary activity of raising horses. After a period of imprisonment and the loss of his horses and all his cash savings, he was told to work in a state butcher shop (*Konsum*) installed on his premises. For years he unsuccessfully petitioned for the right to reopen his business. Attempts at setting up shop in a town in a neighboring district for which he had received the green light from the pertinent local authorities also came to naught. Each time, his nemesis, a *Kreis* authority, would block his path. But then, he was unexpectedly courted by these same authorities to open a butchery with his mothballed equipment for the local LPG. As he explains, "I received a writ asking me to present myself. That was in 1979 or 1980. They told me what they had in mind. The idea was to establish the first LPG butchery. They needed me to accomplish this. It was the same person who had accused me of wrongdoings the year before. I told him that he should inform his collaborators about the actual state of affairs; for during the previous year I had succeeded in making the *Gnadenschiedskomission* (grace commission) in Halle drop the relinquishment of my right to run an independent butchery. I had petitioned the commission twice, and the second time, they conceded my request. But, the suspension of the sentence did not mean that I could engage in my trade again."

With all the potential avenues of establishing himself as an independent butcher again blocked, Mr. Herder had little choice but to agree to the LPG deal. "The manager of our LPG had to appear with me at the town hall and they told him, 'Listen, you and Mr. Herder must open a butchery.' They had asked me before whether I would do it. I had asked for some time to think the matter over, even though I knew that I had no choice, for that's how things were. First of all, they had no one else who could have done it and second, if someone else had come here (and used my facilities) the butchery would have been in a sorry state within a year. That would not have worked at all. Well, let me tell you, the person who had given me so much trouble before, now even held my coat for me. That's how they were. I told them to their face that they merely wanted something from me, but they would not get anything from me. But then they went on and on and he told me that he could, after all, dispose of my machines as he pleased. I told him, 'That's what you think,' put my coat on, and started to leave. They started pleading with me, 'For God's sake' and so on. 'You can't speak to me like that,' I retorted. 'Those are my machines and you are not going to get them.' He continued to plead and wheedle. So I told him, 'Fine, but I don't even know what the *Ratsvorsitzende* (head of the council) will say, for until now he has always been against me.' 'You can speak to him right away,' came the reply. I protested, 'He doesn't have office hours now.' (But that didn't seem to matter.) The man from the *Stasi* did not want to give me time to think about the matter, but the other man said, 'Well, let's give him eight days.'

"After eight days I returned and agreed to do it. I made certain demands and then we started the thing. There were no butchers, so they had to find other people. One was a home butcher, who had originally shorn sheep, another was a home butcher who was trained as a mason, the third one was also a home butcher who had worked in transportation here at the flour mill. Then, we had a woman who had worked in a hair and boar bristle factory in the neighboring town and another woman who had worked in the LPGs kitchen. So we were six or seven persons right from the start. . . .

"The LPG had more than 1,000 pigs and 500 to 600 head of cattle. I had a good relationship to the manager. I did this for him and it worked out well. The operation was profitable and so it was smooth sailing. The money that came in did not belong to me, otherwise I could wheel and deal however I wanted.

"I received rent for my machines and I had a truck—I still have it today; it is now twenty-two years old and is still running—for which I also received rent. Whenever there was the slightest problem, I could bring it to a workshop and it was repaired. That did work. The manager never told me I couldn't do that. I had told him from the start that it had to be a properly equipped repair shop, I didn't want any old mechanic to fiddle around with it. It worked. They always claimed that the provisionment of the people was the first order of business. That helped even on the road. When a road was blocked because of a traffic accident, or something like that—at the time we didn't have cooling equipment—we would ask the police to let us through because people were expecting us, and they would wave us on."

Concomitantly with the transformation of less collectivized cooperatives into "more advanced" forms and with the consolidation into larger units, the share of the profits to be distributed according to the amount of land brought in by the individual members dropped. By the time of the *Wende*, while land was still nominally private property, the so-called *Bodenanteile*, which amounted to 30 to 40 marks per hectare, barely covered land taxes. According to a former LPG manager, the state leased all the land. "The cooperative did not enter into any lease arrangements, but was merely a third party to these arrangements. This was done so that no one could engage in land speculation. In order to protect the land, there was a law in the GDR according to which anyone who touched agricultural land had to obtain official permission and then usually had to pay 400,000 marks per hectare if the land was taken out of cultivation and used for building or other nonagricultural purposes. Today I can obtain a 400,000 premium for leaving a hectare of land fallow!" Also, while the original members were forced to contribute considerable sums in cash, animals, and buildings to enter cooperatives, later, individuals were admitted without having to make any contribution whatsoever. Another former LPG manager commented, "One only had to complete a request form and then the plenum of the cooperative members decided whether the person would be accepted or not." The cooperatives had, for all intents and purposes, been transformed into collectives. The former LPG manager who was, generally, most positively inclined toward GDR practices, saw the move toward opening the LPGs to individuals who made no initial contribution as the most negative development in their history. "That destroyed the cooperatives. It slowly dissolved the inner circle of those who had made a contribution and a cooperative democracy resulted, where those who had not brought anything into the cooperative could rule over those who owned the means of production. For, in time, the latter became older and fewer, while the former, who came later, eventually had all the say. That's what democracy was all about."[4]

The progressive collectivization of GDR agriculture did not take place without resistance on the part of members and managers of cooperatives. In fact, such resistance may have thwarted some of the more extreme experiments of the state, such as the creation of vast agro-industrial complexes. The combination of working within the constraints of the system and resisting a further erosion of the autonomy of cooperatives is exemplified by the description, in his own words, of how an extraordinarily successful manager of a cooperative thwarted attempts to force the cooperative to join in the purchase of machinery and relinquish control over its use to a larger unit. He had been placed at the head of a failing cooperative and had managed to turn it around. Then came a series of attempts at further collectivization. "That was a difficult time . . . every development—the euphemism was 'step in the path of societal development'—had the ultimate goal of dispossessing the cooperative farmer of the private property that he had brought into the cooperative and also of what the cooperatives had accumulated jointly and this, whenever possible, on a voluntary basis. So, whenever a new agricultural policy was announced, I always asked myself what their real intentions were. I would then try to alter the course of events a little. I will give you a practical example. . . . [The manager of the neighboring cooperative] was a stalwart party man and so they chose his cooperative as a model unit. We and [another cooperative] were to be involved [in the experiment] as well. At the time, new kinds of equipment had become available, for example, to harvest sugar beets and a new series of combines. We were supposed to purchase the equipment jointly. We cultivated 5,000 hectares and they had 6,000, [so such a joint purchase] made no sense whatsoever. The equipment was supposed to be stationed at the agro-chemical center. In other words, we were supposed to pay for the machinery, but it was not going to be stationed here. It was supposed to be stationed at an intercompany unit, a so-called support of the working class (*Stützpunkt der Arbeiterklasse*) and managed from there. [Although this happened in 1976] the development toward such cooperative arrangements (*Kooperationen*) already started in 1960.

"Seeing this, I told my principal accountant, a former officer of the navy—he has died since—that we had to prevent this type of joint ownership, for it was the beginning of a new thrust [toward further collectivization]. 'Even if our worst fears prove to be unfounded, the equipment will not be maintained.' He answered, 'Leave it to me. I will impede it. I will quarrel with the other [collective's] principal accountant, so that we will never come to an agreement. I am retiring in one year, so nothing can happen to me, while you still have twenty years of work in front of you.' So we bypassed the problem internally. [Nevertheless], during that same year, they extracted 2.5 million marks to build an agro-chemical center and a drying plant. At the time I told my chief accountant, 'We will never see that money again. God willing, we shall be able to support the loss.' And so it was. We never saw a cent of it. Not even after the *Wende*. It was all gone."

In conclusion, from the perspective of a distanced observer, it might appear that Soviet-style collectives in general, and GDR collectives in particular, ended up looking very much like the fully nationalized enterprises that cultivated some 10% of the land. Indeed, the process of collectivization was characterized by the move from voluntary to involuntary relinquishment of individual and local control. Voluntary affiliation was replaced by coerced affiliation; the option of private animal husbandry was severely curtailed; and, although never abolished, land ownership no longer carried any advantages for the owner.

Furthermore, the autonomy of the collectives themselves was undermined. Successful collectives were arbitrarily joined or forced to share their resources with less successful neighbors. And, as will become apparent in the following chapter, even the larger collectives that resulted from these amalgamations were severely constrained in terms of what they were allowed to produce and could not dispose of their surplus freely, further reducing differentiation between collectives and between collectives and state farms.

Indeed, one of the ways in which the history of agriculture under communism has influenced farmers' identities in the former GDR is in the desire of those who chose to strike out on their own or who have returned from the west to turn back the clock to the time before cooperatives were promoted or even before the land reform of 1945. And yet, it would be a mistake to view the history of agricultural collectivization solely as characterized by the inexorable decline of individual and local autonomy and, consequently, its influence on identity formation solely in terms of the wholesale acceptance or rejection of that process's ultimate goal of transforming individuals into cogs in an overarching socialist bureaucracy. The process can also not simply be viewed in terms of the conversion of the managers of collectives into actors who, like the managers of state farms (and indeed all forms of state enterprises), had to be extraordinarily resourceful because of the deficiencies in the allocation of inputs. Rather, the history of agriculture in the GDR influenced many of the actors, particularly the managers of collectives, in more subtle ways. First, collectivization did not result in the complete nationalization of farming. Managers of collectives were acutely aware of the differences that distinguished them from managers of state farms and made every attempt to maintain these differences. They carried the identities formed by this continued struggle for local autonomy over to the period after the demise of the GDR. They accepted and identified with certain aspects of the process of collectivization and resisted others. Conversely, the state acknowledged the de facto permanence of two potentially conflicting relationships to land.[5]

The grand experiment in land collectivization and consolidation that was intended to equalize access, management, and income and integrate related industries in a democratic and profitable fashion for the provisioning of the people was caught between the divergent interests of its members and the state. These tensions continued even during the period of relative stability of the late 1970s and 1980s when no major efforts at further collectivization, except the furthering of greater coordination between crop and livestock cooperatives, took place.

This chapter has given a brief historical overview of agriculture in the region studied, beginning with "traditional" pre–World War II patterns, then continuing with the land reform after the war and the subsequent collectivization. Each stage was marked by competing models or policies on the part of the state, models that have influenced the identities of the various actors to the present. The day-to-day operation of the cooperatives in the 1970s and 1980s, the manifestations of these tensions, and their effects on the identities of the farm managers are the subject of the following chapter.

CHAPTER FOUR

The LPG

Introduction

During their apogee, between 1973 and their dissolution in 1990–91, the collectives were imposing in their complex organization, size, and range of activities. The size of the former crop collectives studied ranged from 4,000 to 6,500 hectares, with most of them clustered around 6,000, which is close to the GDR average. Each of these crop and fodder cooperatives was associated with two or three livestock collectives, each of which typically raised some 1,000 cows plus other livestock, forming a *Kooperation*. They were the pride of the GDR. In the words of one former LPG manager, "Agriculture and the farmer enjoyed a high status in the GDR. It was, after all, *the* sector that earned foreign exchange. If it had not been for agriculture, the entire economy would have gone down the drain three or four years earlier." "We were mandated to produce more and more," adds another former manager, "but we were also paid well for our products. I don't know of any farmer in the GDR who engaged in some planning who was poor. Farming was an established and integrated entity that could show major accomplishments and thereby advance [the economy]."

The consolidation of collectives into gigantic units described in chapter 3—in addition to leveling differences between more and less successful units—also corresponded to a model to achieve economies of scale borrowed from U.S. industry. Such models, conceived of in the abstract, were applied regardless of considerations of actual productivity. An agricultural

FIGURE 4.1. An LPG cattle barn close to a village near Bernburg.

consultant who was a specialist in firm structure in the GDR explained the failure of the effort to plan LPG size in the following manner: "The original idea was to test large-scale farms to see whether they would work in the East, but the Party took the decision to go ahead [without waiting for the results of the prototype 10,000- to 20,000-hectare farms]. Scientific research in the DDR had come up with an assessment of optimal farm size, but [the advice was ignored] and decisions were made according to political criteria. Already at the time, we came up with an average ideal size of 2,000 to 3,000 hectares. German villages usually had access to about 1,000 hectares of arable land. We figure that a maximum of three villages can be consolidated into one unit. In such a unit, transportation problems are still manageable. Agricultural products are not readily transportable. Beyond certain limits, it is no longer sensible to transport straw or manure, for example. Those limits were recognized in the GDR. Especially the concentration in animal production was totally absurd because of the environmental damage this occasioned. Even the trees in the surrounding area died. These negative developments (*Auswüchse*) resulted from the attempt to impose industrial criteria on agriculture without considering the damage that would result." In addition to exemplifying the process of inexorable enlargement of farm size, the foregoing example also is indicative of the heavy emphasis on what was termed scientific socialism (see Lampland 1995, chapter 5), according to which planning was regarded as a noncontestable scientific endeavor. As Lampland has shown for Hungarian communism, this did not mean that planning was based on scientific studies; indeed, as the above quote shows, such studies were often ignored. Whatever plan was adopted was justified in scientific terms. Outcomes of planning decisions were not subject to falsification since statistics could always be, and mostly were, doctored to correspond to the desired goal.

The Mass Production Model in Agriculture

The GDR economy was characterized by the coexistence of extreme specialization and equally extreme vertical integration. This paradox resulted from the conflict between the government's attempt at instituting idealized organizational forms and its incapacity to effectively micromanage production and distribution. On the one hand, the GDR's emulation of the Fordist model of mass production (see also Voskamp & Wittke 1991) led to an extreme concentration of production and a high degree of specialization. On the other hand, the country's dependence on the Soviet Union; the difficulties of obtaining foreign currency for needed inputs from the West and the channeling of those inputs into defense rather than production of food and consumer goods; and the inability to coordinate production in the different industries forced each economic unit to be as self-sufficient as possible.

In agriculture, this assembly line model of production favored by the government manifested itself in the separation of plant and animal production, even though agricultural experts had warned against such a move. In the area studied, a typical collective specialized in animal production raised 2,200 head of cattle, including 600 cows, 300 sows, and their offspring. Another raised 980 cows as well as bulls, and some 1,500 pigs. A third collective owned some 6,000 head of cattle, including 1,800 cows, and raised some 7,000 pigs. A fourth raised 2,200 head of cattle, 8,000 to 8,500 pigs, 2,200 sheep, and, for a time, some 60,000 young hens. Some showcase farms reached even larger concentrations of animals. Thus one LPG specialized in cattle fattening raised 40,000 animals (Rühle 1991:8); and one gigantic pig farm in Thüringen reached a maximum pig population of 196,000 animals (Grund 1990:24–25)!

The industrial model was also manifested in the specialization within LPGs. Although the large size of the crop collectives made specialization in a particular crop unattractive and they were obligated by law to provide fodder for the livestock collectives in the *Kooperation*, there were specialized units *within* crop collectives (Gollnick et al. 1990:5). Livestock collectives tended to specialize in a particular species.[1] Also, individuals working in collectives often had highly specialized functions. Crop- and livestock-raising were considered separate careers, and individuals specialized further as tractor drivers, hand laborers, agro-chemists, cattle-, pig-, or poultry-raisers, and other specialties (Gollnick et al. 1990:41). In addition, individuals with managerial functions from brigade heads and above were exempted from physical labor. All this specialization should not hide the fact that agriculture in the GDR continued to involve a large amount of manual labor. For example, 90% of the tending of sugar beets was done by hand, and 50% of the women and 25% of the men engaged in heavy manual labor on a daily basis (Gollnick et al. 1990:41).

At the same time that agriculture entailed a considerable degree of specialization, farms were also forced to rely on their own resources to a much greater extent than large farms in the West. LPGs had their own equipment maintenance and construction brigades, which also built homes for the members, crop storage facilities, and (more rarely) distribution, transportation, factories for intermediate and (sometimes) finished goods, and gas stations. This self-sufficiency also extended to the provision of services for the membership. The collectives maintained cafeterias, child care facilities, and vacation resorts for workers. They even built houses for their members. Paradoxically, then, the concentration of specialized functions into fewer units was offset by a fragmentation of supportive functions. As we shall see later, the vertical integration was echoed in the various services in kind and contributions in cash that the collectives were forced to make to the communities in which they were located and even to facilities in the township capital. The boundaries between the collective and the wider community were thus fluid.

The Dynamics of Planning: Rationales and Irrationalities

Let us examine the failure of central planning and the adaptations made by the individual collectives more closely. We contend that in spite of the centralized nature of the GDR economy, or—paradoxically—to a degree because of it, the local responses vis-à-vis centralized management and the procurement of adequate inputs and chronic shortages were quite varied as was the degree of success when measured by the well-being of the members of a collective and of their co-villagers.

The separation of crop growing and livestock raising meant that the animal production cooperatives did not have direct access to land. Rather, they procured feed from neighboring collectives (usually those with which they were formerly joined). The relationship between farms specialized in raising crops and those specialized in raising livestock was fraught with tension. As one consultant explained, "The friction was generally recognized. A lot was recognized, yet nothing could be changed. Someone in the politburo, Comrade Grüneberg—he has died since—got it into his head that this form of production was a good idea, even though it went against the grain of us farmers. I don't know anybody who thinks that it was an optimal solution. But no one could prevent its implementation." Another man expressed his regret about the transformation of the intermediate-sized, integrated collective into giant-sized, specialized ones in even stronger terms: "The 2,000-hectare farm with both livestock and crop production was a marvelous operation. Things improved economically. But in the years after plant and animal production were torn apart, the economy went down the drain."

The rivalry between plant and animal production arose from the ambiguous nature of the dependency between the divisions. On the one hand, their relationship was theoretically open to independent negotiation (although the state provided price guidelines), while, on the other, the prices obtained by the livestock LPGs from the state were fixed. One former LPG manager expressed the resulting tense situation in the following words: "Of course, everybody wanted to negotiate the most advantageous prices, so that, ultimately, we just fought with each other and made no headway at all. For years we were only concerned with ourselves. That was an extremely bad situation." On the other hand, as we shall see later, a relationship of inequality was built into the system from the start, for the state gave official preference to crop growing over livestock raising and dairy farming. The secondary importance of animal production meant that a collective specializing in animal production could never count on obtaining adequate supplies of fodder and sometimes had to send members to the north to buy silo fodder and hay to secure enough feed for winter. At the same time, such collectives never had enough feed with high nutritive value that would have enabled it to increase milk production, for the state feed reserves were always short (Gollnick et al. 1990:53). Thus, instead of feeding a cow a ration of 6 kilos per day, considered the norm today, only half a kilo could be given. The observation of another consultant that plant production, with its inherently better economic prospects, also attracted more motivated and highly skilled individuals, provides an additional reason for the hegemony of plant production over animal production.

Nevertheless, crop and livestock farms did not act as entirely independent agents in their mutual dealings. Their dependence was affirmed by a former crop-growing LPG manager, who said, "We did have to produce a certain amount [of animal fodder]. We were obligated to secure the basic needs of animal production. After all, we were in a planned economy where either the surface to be cultivated (although not down to the last hectare of land) or the amount to be produced were determined from above." In addition, there were councils that coordinated the activities of livestock and crop production. These councils had become increasingly more effective in the 1980s and were given more and more clout in organizing the activities of the member LPGs. Laschewski even speaks about a "reintegration of crop- and livestock collectives into larger organizational units" (1998:102–105), but Gollnick et al. (1990:2) caution that this integration was more ideal than real, for each LPG continued to be an autonomous entity. Such units usually included one crop LPG and two or more livestock LPGs. The latter were represented by an average of five to six members, and the crop LPG, which was larger than the livestock LPGs, was represented by a larger number. The *Kooperation* was usually headed by the crop collective.

The longtime manager of a modern cooperative and former LPG observed, "There was a council of the *Kooperation* of which I was also the head, of which the heads of animal production were also members.

This council had certain sovereign rights. The council was responsible for the global economy of the member collectives. If one of the units had a shortfall, the state did not make up the difference. Rather, the deficit had to be equalized within the confines of the *Kooperation*. This was actually a very good system. It made it possible to intervene in an operation and gave the powers to correct problems. Those [cooperatives in a *Kooperation*] who had the money also had the right to make decisions. I would tell [the managers], 'What you can pay for yourself, you may also decide about. But if you require money, you have to ask first.' That [premise] was accepted and things worked well here. In general, there would be one unit that would have more money. In this area, it was plant production, but in the Altmark, where the land is less fertile, it could also be animal production. The head of the *Kooperation* would generally come from the more prosperous unit. He also tended to be the most skilled manager. So he would make the decisions." The cooperation and equalization among collectives appear to have transcended the *Kooperationen*. One former manager of a collective spoke of the fact that plant collectives would sell feed at lower prices to animal collectives within the *Kreis* (district) that could not make ends meet. As we shall see later, the *Kooperation* could, on occasion, become the basis for the new cooperative structure that replaced the collectives after 1989.

Production Quotas and Export

The reason why plant production was systematically favored over animal production was related to its dominant role in the export economy. A man who worked in livestock raising of an LPG from 1975 to 1990 noted that the official reason given for this preference was that "Plant production farms were given more weight and were preferentially equipped with modern technology and investment capital with the idea that plant production had to run optimally before animal production could be developed. The argument was that one couldn't put the cart in front of the horse. One couldn't build up large animal herds without being able to feed them." But behind the official explanation, there lay another concern, namely, the need to produce for export dictated on the one hand by the agreements between the GDR and other Council for Mutual Economic Assistance (COMECON) countries, particularly the Soviet Union, and, on the other hand, the need for hard currencies to import goods from the West. Grain was the most reliable means of payment and therefore its production was prioritized. In contrast, a crop-producing collective had few incentives to produce fodder, and the grain that was produced and could also have served to feed cattle was slated for export.

The fact that animal products were in lesser demand for export than crops did not mean that livestock cooperatives were exempt from pressures to produce exportable products and could therefore content themselves with adapting their output to the availability of feed. Instead, they were forced to give preference to production suitable for export irrespective of whether or not such an emphasis made sense in terms of profitability. A former LPG manager told us, "Animal production was certainly not efficient. We were forced to keep animals regardless of whether or not they were productive. When the plant production LPG did not produce enough fodder, each animal was given less to eat. But what interested [the government] were the calves: veal, live calves, and even embryo transfers, because there was a demand for them on the world market. Milk was not in demand because even then there were 'butter mountains' in the old *Bundesländer* and the European Community (EC)."

The state did not limit its intervention in agricultural production to the supply and the pricing of intermediate inputs and prices paid to the producer, but went as far as dictating minimum outputs and sometimes even the acreage to be planted in specific crops and the number of animals to be kept. Crop LPGs were given more leeway than livestock and dairy LPGs. While, in theory, the state dictated the acreage an LPG had to assign to particular crops, in practice, this regulation was, in time, ignored, as long as the quota was fulfilled. In contrast, even if a livestock and dairy cooperative could have fulfilled its milk quota with fewer cows, it was not allowed to reduce the size of its herd (and feed the remaining animals more adequately) because the state was interested in maximizing the production of calves. The rigidity of state mandates become obvious in the words of a former manager of a livestock collective, "One couldn't just argue that one could produce more milk with 800 cows than with 950 because there was not enough feed, barn space, and workers. No compromise was possible because [the government] absolutely wanted calves from 950 animals. As a result, milk productivity per cow was half of present levels. In addition, there was an enormous jump in productivity between the end of winter, when the animals had spent many months with insufficient nutrition, and spring, when they could be let out to graze."

The state was also highly inflexible in negotiating production quotas. The same person said, "One could discuss quotas, but one rarely won the argument. . . . They dictated how many kilos of meat, milk, and wool we had to deliver. If we had raised hens they would also have dictated the number of eggs we had to produce. These were not just annual quotas; they were determined for each month. One does have to say, though, that if one was unable to fulfill the quota for a given month it didn't matter too much, as long as one made up for it later. But we were controlled on a monthly basis by the state. Good fulfillment of the quotas was praised, underachievement was criticized. If the plan remained unfulfilled too frequently, a manager could be fired. Such decisions were not made internally but by party headquarters at the district level." Nevertheless, as Gollnick et al. argue (1990:55), "There were considerable differences among collectives regarding the structure of cultivation." For they attempted to optimize production that supplemented production dictated from above. "For example, where there were loess soils, they would plant potatoes that would ripen at different times and different types of vegetables and unusual crops (like herbs)."

The few positive incentives that existed to improve milk production were watered down by the fact that a dairy could not be held to the state mandate to pay a premium for high productivity. If the dairy did not have the means to pay, the premium was simply canceled.

In conclusion, GDR agriculture, as indeed many other sectors or subsectors of the GDR economy, were, in many ways, as dependent on the West as the entire eastern German economy is now. By and large, the dependency between East and West resembled that of the dependency between First World and Third World countries. A flow of cheap grain from the large agricultural units of the East flowed to the West in return for expensive, technologically sophisticated goods. However, unlike the classical North-South dependency (whose nature is presently also far more complex than the stereotypical models), the GDR also met highly specific demands in the West, particularly the FRG,[2] as well as in other COMECON countries, demands that required highly sophisticated production methods. The main difference with the present dependency was that the LPGs could not obtain inputs from the West in any but the most roundabout ways, since all transactions were mediated by the government, whose principal rationale for the exchanges with the West was securing inputs for heavy industry.[3] This meant that the sophisticated outputs from the GDR often had to be produced with inadequate equipment and other inputs.

The Organization of Agricultural Tasks

Central management of agriculture extended to the management of the farming activities themselves, down to timing of the harvest, which led to considerable resentment and, as we shall see later, elaborate avoidance maneuvers.

The consolidation of smaller LPGs into larger ones did give the collectives the financial means to hire specialists who did little else but coordinate the application of herbicides and fertilizer. However, the presence of specialists did not necessarily mean that their knowledge was actually applied. In answer to this question, one former LPG manager said, "You can well imagine that that was not the case. We didn't receive the means. We would call certain fungicides 'gold dust' because they were so rare. Many herbicides could not be obtained at all or only for seedstock. Nevertheless, if you can't do something, but you know that you could do it if you had the means, that's worth something too. At least you are aware of the situation." Given the circumstances, the LPGs would deal with the situation as best they could, for example, by following a strict crop rotation and by combing the weeds from the field with a fine-toothed harrow. Often, some fields would receive all the weed killers, while others would be tilled by hand. Such knowledge could also serve to limit damage to the crops from dangerous products. Considerable risk was involved in the application of some of the available products. For example, some weed killers were known for the damage they could cause the crop if their application was followed by heavy rain, or for their inefficacy when there was no rainfall at all.

At the same time, the large scale of the operations also meant that even the specialists were not familiar with the specific characteristics of particular parcels of land. A man who had worked in an LPG as a specialist in plant protection and now has his own farm explained that "A farmer who plows his own land knows where the soil quality is inferior, where it's drier and where it's wetter. So, when he carries out plant protection measures, he will know exactly which parcel of land has a problem with this or that weed. He will also know what type of wheat to plant in a particular field. None of this played any role in GDR times. When someone sprayed, he would drive over the wheat and sugar beets from morning to evening from March to June without any awareness of the kind of soil he was spraying."

Finally, the large scale of the agricultural units and centralized planning with little or no regard for local conditions often had a disastrous environmental impact. As we have already mentioned, large concentrations of animals in one location increased the cost of transporting the manure to the fields. In order to avoid the cost of building large holding tanks, liquid manure was spread on the fields at any time of the year, increasing air pollution and runoff problems. And, in order to reduce transportation costs and stretch the limited vehicle pool, most of the manure was spread on nearby fields, which accordingly were overfertilized. The ecological damage becomes clear in the words of a former LPG manager: "We sought out a field and spread liquid manure on it during the entire winter and plowed it in early in spring. Then corn was planted there. The corn that grew there became dark green and was so loaded with nitrite that it would affect the cattle. Nitrite cannot be eliminated from the body and harms the physiology of the animals." The problem of too much liquid manure being spread on small surfaces appears to have been common in the entire GDR (see, e.g., Kowalewsky 1990:6) and was a direct result of the high per capita production of animals, its separation from plant production (see Rühle 1991:8), and the high concentration of livestock in gigantic farms. The pollution such farms could occasion is made vivid by the extreme example of a pig farm in the resort area of Neustadt in Thuringia where the pollution created by close to 200,000 animals led to the destruction of 117 hectares of forest where the barns and catchment basin were built. An additional 650 hectares were devastated and 1,800 hectares endangered by pollution. The air pollution by nitrogen was comparable to that of a city of 600,000 inhabitants (Grund 1990).[4]

A major reason for the disastrous agricultural practices in the GDR was the systematic silencing of opinions that were not in accord with official opinion. As Gollnick et al. (1990:73) explain, "The professional press only published articles approved by (the central committees of the SED). Therefore, a public debate was excluded a priori. The authorities preferred nationwide homogenous solutions to the needed multiplicity of solutions because the former were easier to understand and administer. Economists working in universities could only test out their ideas if they had good personal contacts with managers of collectives. Undesirable whistle blowers who pointed to problems and mistakes in public presentations were punished (public self criticism, party punishment) or fired. . . . As a result, there was little criticism of erroneous agricultural policies and the economists contented themselves with measures to contain the damage." Nevertheless, new theories about agriculture did, on occasion, reach the party authorities and were adopted by them. Thus, because of the increasing scarcity of chemical fertilizer, a ban on criticism of the use of large amounts of such products gave way to a theory that focused on the circulation of nutrients within a balanced system of plants and animals (Gollnick et al. 1990:79).

Local Responses

The inflexibility of the demands placed on the collectives and the dubious merit of many of the directives forced the LPGs to find devious means of circumventing them. As Gollnick et al. (1990:38) explain, they maintained one type of organizational structure for official purposes and another one for their own purposes. The stronger the cooperative, the more likely it was that it could get away with following its own devices. In contrast, weak cooperatives had to follow directions from the party, even when it was to their detriment. One former LPG manager recalls: "We had to be very inventive to bend things in our favor. People used to say that there were three levels of lies: a lie of convenience, a true lie, and statistics. Unfortunately that was the way it was in the GDR, the statistics were not accurate. They couldn't possibly have been correct. Even at the bottom, we furnished totally wrong figures and these were passed on to the next level. . . . For example, we were forced to keep 950 cows, although we only had 930 stalls. Money to make new investments was scarce and even if money was available it was difficult to obtain workers. Everything had to be requested through petitions. When one wanted to build a new barn, the capacity was never adequate. For many years we tried to [obtain official permission] to keep only 900 instead of 950 cows, but we never succeeded. But we were inventive; we kept only 900 anyway and declared 950. Someone rarely came to count them and if they did we could always say that the shortfall was temporary. Of course this subterfuge had other consequences. One also had to deliver milk and the state could figure out how much milk was produced per cow. Milk production was generally very low in the GDR anyway—an average of only 4,000 liters per cow per annum. If a farm only produced an average of 3,000 or 3,400, someone would inquire about the low rate. We were able to deal with that problem in one way or another. One had to be astute so that things could continue to function somehow."[5]

LPG managers also found ways to ignore and circumvent the forced district-wide coordination of agricultural activities. One former LPG manager described how he learned to cope with the ridiculous demand to begin the harvest at a specific date as follows: "They had written in the newspaper that such and such [an LPG] had begun to thresh and this would mean that we would have to start too. Otherwise the manager would be called in and he would be asked why he hadn't begun to thresh yet. [So he had to start even when the crop wasn't ready yet]. Some farms lie lower than others. That can't be helped. But when those who were in the most advantageous locations halfway up a hill began to thresh, then those whose fields lay [higher up] in the hills and those with low-lying land also had to start. It didn't matter if the crop was not ripe yet or the soil was soggy; the newspaper said that these and these LPGs had begun to harvest and everybody in the GDR went crazy. At first, you get terribly upset about such a stupid practice because you know better. But then you get used to it and are amused, and you continue to act according to how you have been taught. You no longer care a hoot and say: 'Let them say what they want.' So you may be punished by the party or given a reprimand and then, when new tractors are distributed, you don't get one. Things like that could happen. But, in general, you became thick-skinned. You see, if someone tells you that you absolutely have to begin to thresh, then you simply mow a bit. You begin at the front and mow the edge and everything is fine. That's all you have to do. I remember once we threshed while it rained. We worked in the rain with our threshing combines to the sound of the firemen's band because the council chairman decided that it was time to begin with the harvest. I stood at the edge of the field and cried with rage. But later I laughed about it and thought to myself, 'Let them command. We will do it when we decide the time has come.'"

Another way for an LPG or part of an LPG to gain greater independence from outside interference was to engage in an unorthodox (i.e., not readily classifiable) activity. The manager of a factory that produced oils from herbs explained this strategy in the following words: "The first question government officials are asked when they take their qualifying examinations is how they would handle a procedure. The correct answer is, 'I examine whether it is in my area of competence or whether I may pass it on to someone else.' When we first developed [the process of extracting oil from herbs], the first question that arose was which ministry should handle it. The quarrel over this issue of competence ultimately had the consequence that we were among the very few farms in the GDR that could freely make decisions about land that had been allocated to us. We developed our own technology, set our own prices, and requested the government department that set prices to ratify them. We negotiated contracts and sold on our own. In short, we became independent agents. All of a sudden, we were at the same time a socialist operation subject to a planned economy and the occupants of a small niche in which we could engage in 'private socialist production.'"

Similarly, a part of another LPG that also produced herbs was able to feed seven families from 90 hectares of land that these families had originally brought into the collective, one of the few to have retained the right for its members to maintain private animal production. While they were not allowed to dispose freely of their income, they could at least deposit their earnings into their own bank account.

Inconsistencies in the Distribution of Farm Equipment

LPGs were dependent on the centrally planned and controlled allocation of farm equipment. Tractors could not simply be purchased on the open market; they had to be acquired through formal requests to the authorities. As one manager of a farming cooperative put it, "The tractors had an average age of six to seven years, even though new ones were added from time to time. The allocation of tractors was planned. A request had to be submitted during the so-called *Abstimmung über den Plan* ("Harmonizing the Plan") with the Department of Agriculture of the District Council. Everybody wanted to obtain as many new ones as possible, but only a limited number came into the district. Those were then distributed. The managers of all the collectives were present at this meeting and tried to balance their interests. Of course, this did not happen peacefully. When there is a struggle over scarce resources everybody tries to push his own interests."

The chronic shortage of tractors was compounded by the quirks of the international division of labor among COMECON countries. Russia did not allow Eastern Germany to manufacture high-powered tractors. They had to be imported from Russia and were in even shorter supply than the nationally produced ones.[6]

The next best alternative to obtaining a new tractor was to have an old tractor remanufactured. A former manager of a LPG tractor station and repair shop had this to say about the complicated process: "Let's take the example of a combine. The first year after its purchase and the second it would run with little or no repairs. But before the third harvest it would already have to be brought in for a total overhaul. There was a specialized firm for this. They would disassemble the entire machine until only the body remained. [Then the parts] would be repaired. [But when the machine was reassembled it was done with different parts.] Thus when a threshing drum was taken out of a machine the part was not put back in. [When a machine came back from the factory] it would have a new (i.e., different) motor, new exterior, generator, and starter engine. The manufacturer's name was always written in large letters on the metal housing parts. But when the machine came back from the repair shop it had different names on the housing, because even the parts that composed the body had been switched. Say, they had to weld new hinges onto a part. By the time a new hinge had been installed, another combine would have reached the front [of the assembly line] and the part would be mounted on that machine. The serial number of the machine had not changed, but in reality it was no longer the same machine at all. Only the chassis remained the same. A piece of equipment practically existed only on paper.[7]

"As a consequence, the equipment was different every year. One could be lucky or unlucky. For example, I knew that in order for a particular machine to work properly I would have to make certain adjustments. But when the machine came back from the shop the adjustments were always wrong. The parts were right and the quality and security were assured, but it was no longer the original equipment. I might have the innards of a newer machine, or perhaps also of an older one. After all, parts were scarce, and so many were repaired rather than replaced with new ones. Or, old and new ones were joined together, which, for example, with sprocket wheels, is fundamentally wrong. If I put a harder and a softer cog wheel together the wear and tear is much worse. So, often, instead of the intended repair, more entirely unnecessary damage resulted. It would have been cheaper just to replace the entire assembly rather than to attempt to repair it."

The need for extensive facilities to repair and maintain equipment was thus, as it were, "programmed" into the economic system. With the chronic shortage of raw materials, the life of each piece of equipment and each part from which it was assembled had to be extended to the absolute maximum. The frequent breakdown of aging equipment, the chronic shortages of spare parts, the difficulty of having repairs made and of procuring outside construction crews, and the long waiting lists for new equipment forced each LPG to establish elaborate service facilities and hire its own construction crews, repairmen, electricians, and other service personnel. Not surprisingly, this state of affairs exacted a heavy toll on productivity. For example, a SED party document reported that in the late 1970s combines were in operation only 35% to 40% of the time (Gollnick et al. 1990:50).

In turn, managing a machine pool and ancillary services required special organizational talents and interpersonal skills. A manager had to anticipate needs for parts and equipment at least one year and sometimes many years in advance. As our workshop manager informant explained, "One couldn't just order something today and have it tomorrow. One had to know that one would need something during the following summer and remember that one would not be able to obtain it then because in summer everybody would be looking for the same thing and it would therefore be unavailable. I would have to buy it during the previous winter or even in fall. It was best to go right after the end of the harvest to buy parts for the combines. Similarly, if one knew that one would need tools, welding equipment, or a power drill, one would have to buy it when it was available. A manager was not expected to find a quick way to repair a broken machine, so that it would run again, but rather to have a long-term strategy for production. He had to write orders with six copies. [For example], I wrote an order for a machine tool in 1985 and received a confirmation notice that it woud be filled in 1996. Those were different priorities than the ones we have today."

The chronic shortages led to a constant search for spare parts, which were opportunistically accumulated even if they did not fill any particular immediate need, for they could always be used to buffer possible future shortages or to barter for needed items.[8] This practice was characteristic of what Kornai (1992) calls a shortage economy, where the scarcities themselves resulted in large part from the fact that the system of allocation was so inefficient that all actors attempted to obtain as many inputs as they possibly could, hoarding everything from spare parts to workers so that they would not be caught short when need arose and they could not obtain inputs or get work accomplished in time without such reserves. The same former tractor pool head commented, "Our major concern was to organize the acquisition of things that we didn't need, but knew that we could use to barter for something else. We had two persons who did nothing else but travel around to 'organize' the procurement of things we badly needed on the basis of this kind of barter. For example, one couldn't obtain cars at all and car parts only with difficulty. The LPG's cars had to run somehow, but it was difficult to do anything about it." Under these circumstances, personal connections with managers of other LPGs was of vital importance. "When I phoned someone in the neighboring LPG, I knew his wife's name, how many children he had—things about his familial life—because we would visit one another after work or meet during training programs where we would sometimes be together for an entire week in some place. As a result, the interpersonal ties were very strong and such contacts could also come in very handy."

A repair shop manager and his workers had to be flexible and skilled in making do with whatever was available. One LPG manager stressed, "We did a lot of things ourselves. We even manufactured parts that were unavailable. We had our own shop equipped with lathes and milling machines, all things which a repair shop for agricultural equipment no longer requires today."

Finally, if all else failed, the dearth of equipment could also be remedied by more traditional means. One member of a former LPG recalls how each member of the collective would be allotted one or two *Morgen*[9] of land to weed and to thin out the plants by hand.

In conclusion, the lower productivity of GDR farms compared with western German farms was, as Hartmann (1995:21) points out, not solely attributable to the organizational structure of the farms but had a lot to do with the lack of machinery and spare parts as well as pesticides and other inputs.

Showcase Projects

While centralized control of the access to inputs usually led to chronic scarcity, it could also lead to the diametric opposite: temporary abundance. The state could impose new ventures on LPGs. LPGs could be forced to engage in ventures against their wishes. The following case is an example of a plan decided upon by the party and imposed on the District Council (*Rat des Kreises*), which then forced it upon an LPG that had to borrow heavily to fulfill the plan. One former collective manager told us, "We had to build a barn for sheep and acquire sheep because the leaders of the Department of Agriculture of the *Bezirk* decided to do it and instructed the *Kreis* to carry it out. While it is true that sheep are suitable for the area, our LPG didn't have any desire to raise sheep. However, we had to build a barn, purchase sheep, find a shepherd and so on, anyway. It was in the plan and was not a subject for discussion."

Such centrally planned projects also included ill-conceived or economically questionable pet projects of the *Rat des Kreises* or some higher-ranking party official, reminiscent of the showcase projects in Russia under Stalin.

One could argue that large, flashy projects are particularly attractive in planned economies not just because of the prestige accrued by flashiness per se, but because the centralized control over the allocation of scarce resources makes at least their conception and the initial phases of their realization imminently feasible. Problems with such projects often emerge only after the enthusiasm that provided the initial impetus for their realization has waned, that is, when the time comes to maintain or even to complete them, which requires continuous longer-term feedback between different levels of the power hierarchy.

Sometimes the idea of initiating a showcase project originated in an entirely unrelated activity. An example of such ad hoc or opportunistic decision-making was the construction of an ambitious irrigation project born out of the ashes of an industrial project that had failed. A former LPG manager recounts, "Once we constructed a gigantic irrigation system, more or less because the state forced us. The system was capable of watering 2,700 hectares of land, a mammoth project that would simply not be feasible under present conditions. Pumping the water is simply too expensive. The system used filtered water from the Saale River. Wells were built along the Saale and the water was filtered by means of a deep gravel bed to make it usable for irrigation purposes.[10] The origin of the project was totally unrelated. The wells had been dug in order to expand the potash factory. After the first ten wells had been dug, the original project had been abandoned. Subsequently, fifty additional wells were dug, a pipeline was laid to the top of the hill up here, and a large water reservoir in the form of a pond with a 15,000 cubic meter capacity was built where the water was also warmed. From there, the water was carried to the fields by means of an underground network of pipes. The project cost several million marks."

Ambitious projects were, on occasion, initiated from below. If an LPG manager was well connected in the bureaucracy, he could take advantage of the party's penchant for showcase projects. One former LPG manager who conceived and built a distillation plant to produce plant extracts conceded, "To be honest, I have to tell you that I was only able to build this installation because I had the necessary support. I had a brother who was the technical director in B. He could assist me with the technical aspects of the project. I had connections which were as crucial then as they are now. I had been in the army and I still had a number of colleagues from those times who were in the ministries and places like that who gave me the go-ahead. So, for example, I was able to obtain 15 tons of stainless steel for our plant which during GDR times was as scarce as gold." He claimed that he could get his ideas accepted by the *Kreis* council as long as they could wash their hands of all responsibility. "The chief of the party at the level of the *Kreis* wanted to shine too, after all. So he would say, 'Fine, go ahead and do it, but if something goes wrong, I don't know anything about it.'"

Even research and development (or what should have been research and development) was subject to the same ideology, according to which only flashy, large-scale experiments were acceptable. For example, after some initial rudimentary experimentation with a square metal box soldered onto an ordinary stove to distill oil from herbs, the LPG in the example described previously was directed to move immediately to build a very large-scale device. The result was disastrous. As the former manager explained, "We realized that we had fallen into the trap of dogmatism. It was all wrong. We had built a distillation device of I don't know how many meters in height and width and immediately put in tons [of herbs]. We did solve the problem of distillation, but no one knew how to get the residues out again. After making a few additional tests and crawling around in the device like coal miners to hack and stamp the residues out, we had to admit that it couldn't go on like this. So we sat down once more and found that we had to maintain a particular maximum diameter and height so that it would empty itself on its own. . . . The [device we came up with] served us well then and continues to do so at present."

Communal Tasks

The LPGs played a major role in funding projects in the surrounding cities, towns, and villages, a function that was legally mandated (see Laschewski 1998:37). They constructed and maintained roads on LPG land, which were often open to the public as well,[11] and funded the construction of kindergartens, halls for community events, music pavilions, hotels, facilities for dentists, stores, sports centers, water supplies, and even entire apartment buildings. To this end, they often used their own construction crews and equipment. They were asked to pay for medical equipment for the *Kreis* hospital. They were major contributors to the annual budgets of sports clubs. Finally, they made substantial contributions for foreign development projects. "The collectives were the entities that still had money at their disposal for things like that," a former LPG manager said. "A good mayor would always have three or four LPGs that would contribute to public projects, even if it was just 50,000 or 100,000 marks. But it varied. Some mayors did a lot and others didn't do a thing."

More than one community could benefit from LPG contributions.[12] The overall average annual contribution of one LPG amounted to between 300,000 and 400,000 marks; a second LPG figured an average of between 40,000 and 50,000 marks, and a third, a total contribution of 1.8 million in a period of ten years to the three communities located within its perimeter.

The LPGs' seeming largesse was, to a large extent, more apparent than real. The sums contributed to the community could be deducted from tax liabilities to the state for the year following the one in which the expenditures were incurred.[13] Therefore, although they made for a less impressive balance sheet, the expenditures did not necessarily hurt the LPGs' economy. However, LPGs would often take up loans to pay for equipment and barns while, at the same time, making impressive gifts to the community that was not allowed to borrow money. While interest rates were often as low as 1%, and interest did not, therefore, constitute a major burden for the LPGs at the time, these debts have, as we shall see later, come to haunt the cooperatives that developed out of the collectives after the *Wende*.

While the LPGs presumably did not have to worry about obtaining the necessary inputs for most of the community projects they were asked to fund, they did have to scramble to obtain materials to maintain the roads. For example, one LPG bought coarse chalk stone gravel, which was a by-product of soda produced in a nearby factory. Since it was a waste product, the sale of the gravel was not subject to regulated allotment. Thus know-how and planning were important in communal tasks just as in internal ones.

Capital Investments and Debt

In a planned economy, an economic unit's capacity to pay for capital investments and, indeed, to a degree, even to be able to cover its operating expenses is less important than the availability and political allocation of resources. Indeed, the meaning of money is different from the one in a capitalist economy. Thus, although money continued to serve as an accounting device, and—to a degree—indicated the economic health of an enterprise, it constituted but a secondary predictor of an economic unit's ability to obtain productive inputs.

If an economic unit had a surplus, most of it—two-thirds in the case of LPGs—was taxed away and an LPG's profits could not (as we shall see later) be readily distributed among the members. Conversely, the consequences of incurring a debt were not the same as in a capitalist economy. As we have already noted, interest rates were low, so incurring a debt, once approved by the corresponding authorities, had little impact on the lives of the members. Neither was there, according to some informants, much choice in the matter. The state's rationale for appropriating surplus and for treating a collective's need for credit and other factors of production independently of its means of paying for them followed the same logic. The state considered surplus to be a direct result of its pricing policies, which were considered generous. Hence the state felt that it had a legitimate claim to surplus. In addition, since in the 1960s workers in agriculture were earning far more than industrial workers (5 marks versus 3 marks per hour for the latter), and were exempt from taxation, the levies were a means of leveling the differences. The manager of a former LPG that never had to borrow money expressed his frustration as follows:

"We always had enough money here in M. Neither are we carrying any old debts, or new ones for that matter. We always paid for everything in cash. We would have had the means to buy all sorts of things in GDR times, but everything was preallotted (*bilanziert*). If I wanted a new threshing combine, I had to sign up and wait my turn on the plan. Only then could I make the request. Let's say that ten threshing combines had entered the *Kreis* and fifteen requests had been made. Then the *Kreis* administration would decide who would receive one and who would not. That was the best-case scenario. Usually there were fifteen requests and only two or three threshing combines would be available. But it didn't matter whether [an LPG that had made a request] had the money to buy the machine or not. If an LPG did not have the necessary cash, it received a low-interest loan from the state. The decision was made according to other criteria. They would calculate how many hectares each collective had to thresh and how many threshing combines it already had [and allocate the machines accordingly]." The scarcity of inputs combined with the fact that allocation was made by nonspecialists who based their decisions on political considerations was, according to the same informant, the worst aspect of running a farm during communist times and shows the relative unimportance of money.

The rationale for incurring debt is clarified by the comment of another LPG head: "It was, after all, impossible to avoid taking up a loan. There was no cash. The LPG had to give all profit to the state. So when something had to be built, it was forced to take up a loan. Every year, amortization and interest were paid which could be deducted as a cost and would not have to be delivered to the state. In other words, everything turned in onto itself."

As the following example shows, some LPGs accumulated large debts as a result of inefficient operation because of the inadequate supervision of the state. "The LPG 'N' was an 'LPG with a long name' (i.e., a bad reputation), which meant that it was economically weak. The LPG had incurred debts. In particular, it had to borrow to balance its current account and bridge shortfalls. The LPG did not produce enough. No one examined whether the crop farm served the livestock and dairy unit adequately. A number of studies were conducted, including a doctoral dissertation. They all stated that the LPG set aside 3,200 square meters per head of cattle, which did not sound too bad. But when one figured the production in the amount of feed produced, it amounted to only 70% of what other LPGs specialized in growing crops could sell to their animal production counterparts. No one paid any heed to *that* fact. The officially acceptable range was 3,200 to 3,800 square meters so it didn't matter [that the production was actually deficient]." The shortfall was just borrowed.

It should be noted that the view that LPGs who incurred large debts were not necessarily much worse off than those that did not was not universally shared. The LPGs could, after all, keep one-third of the profit they made. As a result, some LPGs managed to stash away considerable sums of money. While they could not freely dispose of these balances, in part because of the obligations LPGs had to finance community projects, and in part because they could not obtain the necessary materials and equipment anyway, some LPGs did, as we shall see later, find ways to distribute profits to their membership at higher than usual rates.

Compensation

Summarizing the economic standing of LPG farmers and more generally of workers in the GDR, one LPG manager said, "The LPG had a lot of advantages. There was a regular work schedule, which farmers usually don't have. In the GDR earnings were relatively low, but there were many social benefits." Indeed, consultants had the highest praise for the free child care facilities and the free medical care provided by the GDR system and the fact that working mothers received a twenty-six-month maternity leave with full pay for the first child and a full year for additional children. They could extend this leave for two extra years and still receive partial pay. They were guaranteed the same job or a similar one upon their return (Dodds & Thompson 1994:11). They enjoyed the "household day," a free day each month for all married women and single women with children under age 16 as well as women without children who had reached age 40 to engage in household duties; time off to take care of sick children; child subsidies; and a marriage benefit (Dodds & Thompson 1994:13). In addition, families who had been subjected to sudden misfortune could obtain assistance from the LPGs' social fund. Schools were free and so were kindergartens and nurseries. "Communal meals were cheap in our enterprise," explained the same consultant. "A complete meal cost only between 0.50 and 0.80 marks. (In fact, even retired workers could eat lunch at these rates, which were heavily subsidized from the LPGs' social fund.) There were no unemployed and no one was without an apartment. Even alcoholics were taken care of. Sometimes things were almost too humane. Some people didn't appreciate it and trampled on it. Today no one pays any attention to such cases. If someone doesn't want to do anything, he has to bear the consequences. Such things didn't exist at the time, so not everything was bad. In the last years before the *Wende*, we earned an average wage of 5 marks, that is, 1,000 marks a month net. That doesn't sound like much, but [life was cheap]. A new apartment with toilet, bath, and hot water rented for 50 to a maximum of 80 marks."

LPGs had more flexibility in determining wage levels and in deciding what percentage of the remuneration they would pay in the form of monthly wages and what percentage in the form of annual distributions of profit than did state enterprises (including those in agriculture), where wages were determined according to fixed tariffs. Each LPG was allowed to establish a wage fund that it could disburse freely. Workers were paid a wage that varied according to the number of hours invested and the type of work. Women and men received the same wages for the same type of work, except in administration, where women were not paid as well as men. Nevertheless, women tended to cluster in the less well paid categories. For example, they were more likely to be hired on livestock and dairy farms, which, as we have seen in chapter 3, had lower revenues, and the work was therefore less well remunerated, than on crop farms. The inequality in pay was particularly evident in families where women and men were placed on separate farms when livestock and crop production were first split up. Livestock care was seen as women's work and tractor driving as men's work, and so the split in production types also led to the separation of the workforce by gender.

There were gender differences in remuneration and working conditions even within the same LPG. In addition to receiving lower pay in administration, women were also more likely to be relegated to do the dirtier work, like cleaning pigstyes. They also engaged in those activities requiring intensive hand labor, like hoeing sugar beet fields by hand. Indeed, as we have already noted, women were twice as likely to engage in physical labor on a daily basis than men (Gollnick et al. 1990:41). Much of the seasonal after-hours work was undertaken by women. For example, women would work an extra four hours during the onion harvest preparing onions for shipment. However, some women did engage in jobs that required technical skills. They tested milk, worked in experimental labs, and led brigades, and on occasion, even LPGs.

The ratio between the lowest- and the highest-paid worker could amount to as much as 60:100 or 55:100. A manager was paid about twice what a top-level tractor driver could earn. In contrast, a brigade leader often earned less than a tractor driver, giving him or her few incentives to make sure that the more productive workers received the highest remuneration (Gollnick et al. 1990:83). However, if the LPG paid an average of above 12,000 marks per year per worker, it had to pay taxes amounting to 90% of the excess amount paid out.[14] As a result, collectives attempted to use tricks like giving gifts of watches and other items during state holidays and maintaining travel funds. Thus members of the LPG might receive free potatoes or wheat, or straw for the animals they raised at home. Some LPGs managed to pay out even more of their profits through clever accounting practices. When we expressed surprise that during the last years before the *Wende* an LPG manager paid bonuses amounting to between 25% and 35% of annual wages, he replied, "You *would* like to know how we did that, wouldn't you? Well, we did our accounting on a computer at a computing center. We engaged in a lot of specialized cultivation that entailed manual labor. We figured the number of hours according to a book. But that book was actually already ten years old. Or we did it according to calculations that had been made in 1955, which figured seven days to weed a hectare of sugar beets, and so on. We fed all that information into the computer and the computer would tell us that 500 hectares of sugar beets entailed so many hours of hand labor and hence so many units of full-time work (*Vollbeschäftigte Einheiten* [VBE]), so many hectares of onions entailed so many VBE, and so many hectares of marjoram entailed so many VBE, and so forth. And so we always had fifty or sixty more workers in the computer than we actually employed. That money went into profit distribution. That's how we did it. We didn't engage in any form of tax evasion, nor did we fail to follow governmental dictates. I had also already figured out what I would do if someone found out what I was doing. And in one way or another we were always lucky—almost all the LPGs around here were lucky. In addition, we made distributions in kind."[15] Since plant production was favored over animal production, annual bonuses could be as much as five to six times higher in the former.

Although a minimum income was assured, some workers worked outside the farm during periods of low labor demand. One consultant, who had worked in an LPG adjacent to a city, said that there was not enough work on the collective in winter. He explained that while a worker could work up to 120 work units per month (at 10 to 12 marks per unit) during the peak threshing period, more average conditions brought 70 to 80 units and in winter 50 to 60 units were the norm. Therefore, he and other members of the LPG had to work in industry during that season.

Although some of the former LPG managers spoke nostalgically about the safety net provided by the communist system, they also decried the lack of personal incentives to work hard. One former LPG manager answered the question of whether the LPG had to carry a lot of dead wood as follows, "Yes, we couldn't get rid of anybody. It could happen, that we fired someone for alcoholism and failure to appear at work, but then they couldn't be placed anywhere else, and so we got them back again. I believe that we stood our ground and refused to take such individuals back more than twice. When we were threatened with a fine—not the LPG, but the manager personally—I asked how much the fine would be. They answered, '800 marks.' I replied, 'Fine, send us the bill and keep the man.' None was sent, but it could have led to a lawsuit by the labor court. . . . Of the 200 workers we employed, about 100 worked very hard and very well; another 50 constituted the weak average; and the final 50 were dead wood we had to carry along year after year. Perhaps the figures aren't accurate, but it was more or less like that. They were present at work but, if you compare them with the second group, they were worth half as much and only 10% of what the first group was worth. But don't misunderstand me, there were always people who worked very hard, came on time, and worked intelligently. Of those, 50 are now in early retirement and the other 50 are still working for us."

While wages were sufficient for the basics of life, if a family wished to purchase anything beyond the basic consumer goods, household savings could be rapidly depleted. As the informant quoted above explained, "Electronic equipment was horrendously expensive. A small radio, worth 100 DM today, cost 2,000 marks. A basic television set without remote control cost 6,900 marks. Today you can buy one for 500 DM. So, with an average monthly wage of 1,000 marks, the savings of an entire year would go into buying that set." As a result, individual farm working families would, in addition to their normal work, take on the care of a hectare of land which they had to cultivate by hand during their free time. This brought in an extra 1,000 marks, which was spent on vacations or to save up for a new *Trabi*.[16]

According to our consultants, most agricultural workers (one farmer who is now independent estimated 80%) owned their homes. Often, the LPG would contribute 10,000 marks to the construction of a house if the owner had worked for the LPG for fifteen years. The LPG also made gifts to newly married couples. This did not, however, necessarily reduce their expenses for housing. In the GDR, rent was heavily subsidized by the state. In fact, it did not cover maintenance costs. The only reason privately owned housing was available was state regulations that forced owners to take in renters if their homes exceeded a limit set per household member. Needless to say, an owner had little incentive to maintain such properties. The ideal was to own a small house, where one would not be forced to lodge renters and which one could manage to maintain.

Homeowners faced an uphill battle to procure supplies, including long delays, waiting in line at government stores, and elaborate bartering. As one former LPG worker described the situation, "All hell broke loose already if you wanted a piece of lumber. If you built a new house you might get 2 or 3 square meters of lumber. People would stand in line all night just to obtain a permit to buy 2 or 3 square meters of wood. The same thing was true for a bag of cement. For that you also had to stand in line for almost an entire night. . . . [Sometimes] the first persons in line would stumble into the store drunk [by the time it opened]." The crucial links to obtain scarce resources by means of the connections made at work were women. For example, the wife of one man who was in charge of the tractor pool of an LPG worked for a roofing company and could obtain roof tiles for the home of the extended family that also housed their private farming operation.

All in all, farming was remunerated as well or better than other sectors. After an initial disadvantage, subsidies enabled farmers to enjoy a relative standard of living that many of them can no longer uphold today. These subsidies took the form of price guarantees and supports for farmers located in regions with poorer conditions and for activities like raising sheep, which could not have competed on the world market but which were deemed necessary for the self-sufficiency of the GDR (see Luft 1997:12). As we shall see, the sudden removal of these supports—although many of them were replaced by new subsidies—contributed to the initial dislocations after the *Wende*.

Supplementing Wages: Individual Production by LPG Members

Members of collectives could, and very frequently did, supplement income from working for the cooperative with individual production. As we have already indicated in chapter 3, LPG members were allowed to cultivate half a hectare of land for themselves. By 1977, landless farm workers also were given the right to such land. As a result, by the early 1980s, some two-thirds of the members of collectives also produced privately (Hohman 1984:17). A handful of farmers were able to increase this amount by cultivating marginal lands. Thus the authorities "discovered" some 23,300 hectares of land that was not under cultivation in 1982, some of which was given to the LPGs to cultivate, but some of which was assigned to small-scale producers (Hohman 1984:18). Even road ditches could be assigned for cultivation.

The GDR's policy regarding individual productive activities by LPG members was characterized by ambiguity. While such activities had little support in the 1960s, the resulting decrease in individual production led the authorities to change their policy (Gollnick et al. 1990:4). Larger-scale private ventures continued to be discouraged, but smaller-scale production was actively encouraged. By 1978, members of collectives, farm workers, retired workers, and industrial workers who farmed on the side could earn up to 7,000 marks per year from private agricultural activities without paying taxes (Hohman 1984:16). By 1989, 33% of fruit production, 13.5% of vegetables, 33% of eggs, 25% of rabbit meat, 15% of pork, 29% of wool, and 98% of honey was produced in this manner (Gollnick et al. 1990:4). The state controlled the size of private operations by restricting direct access to land to an absolute minimum. More specifically, it directed the nature of the private operations through its control of inputs. Thus, it entered agreements with individuals to fatten pigs and bullocks for which it furnished the feed (Hohman 1984). Finally, it influenced private production by the nature of price supports for staples. Because of these subsidies, it was often to the producers' advantage to sell to the state rather than directly to the consumer. In fact, sometimes it was more advantageous for individuals to sell their production to the state and supply their own needs from the state store rather than to consume their own produce.

Inputs were obtained in a number of ways. First, at least in some collectives, each LPG member, including the managers, was obligated to weed one-quarter of a hectare of beets by hand because there were always weeds left after the machines had gone through and weed killers were in short supply. In return, the worker had a right to 20 hundredweights of grain. This amount could be increased by taking over the obligations of other workers. For example, one former LPG member and his father weeded 5 hectares of land. He said, "On Whitsunday, when everyone else took off, we would stand in the middle of a beet field hacking away like world champions." Chicken feed could be obtained from the state cooperative that purchased the eggs. A producer would receive 1 kilo of grain for every three or four eggs delivered at a price that depended on the season. Some persons would act as intermediaries and were paid on a consignment basis for collecting the eggs and delivering them to the state store. However, individuals engaging in such semiprivate activities full-time forfeited pension benefits. Those who raised pigs and fattened bullocks also had a right to purchase a certain amount of grain (350 kilograms of corn per bullock and 30 kilograms of food concentrate per suckling pig). Typically, small-scale producers raised two or three pigs, sheep, goats, or bullocks, but even some families with no direct access to land managed to keep larger numbers. Thus, one family, who has since established a larger private farm, raised an average of four bullocks, twenty to thirty pigs, and 200 hens. East Germans, particularly women, whether LPG members or city dwellers, also managed to produce large amounts of vegetables and other produce on small garden plots at the edge of towns and cities.[17]

A very small number of individuals succeeded in gaining access to more land and in running larger operations. Although their every move was carefully observed, and they were subject to frequent intimidation, their activities were grudgingly tolerated, provided that they could prove that they were only cultivating bits and pieces of land that would otherwise have gone to waste. The operation of one such farmer, Mr. Baumann, was located in the middle of a city. There, the family raised seven bullocks, some twenty pigs, three goats, and a horse or two, as well as some 150 hens. Although their own land was incorporated into the LPG where Mr. Baumann also worked, he, his wife and two sons cultivated some fifteen to twenty plots amounting to 9 to 10 hectares of land. They had managed to obtain the land from urban households with no interest in cultivating their garden plots. Mrs. Baumann engaged in farming and egg-selling activities on a full-time basis. She brought the eggs to the state grocery store. She explained, "That was quite a distance, but in those times, 1965–66, when the children were small, I always got a lot of bananas. Often, I spent the entire amount I earned from my eggs for bananas and other fruit, but at least one had a little something for the children. Oranges were also very scarce. They were usually only available at Christmastime. For three or four eggs, depending on the season, one had the right to purchase 1 kilo of feed at half of the usual price. We also received corn. All of this was handled by the *Konsum*. My mother, who lived at the time, also helped a lot. We had to weed and thin the sugar beets all on our knees. And yet, it still was fun." Even a woman renter helped in the cow barn. They managed to acquire a small, secondhand threshing machine and an ancient tractor that they brought back to life. "We had two fillies and a combine," Mrs. Baumann continued. "They wanted to know where we had obtained the combine and how much we had paid for it. I answered, 'Since you know where we got it, you must also know what we paid for it.' I was at home alone and they came with the veterinarian. I was not astute in such matters. I became angry and practically threw them out of the house. But then my husband had to appear in front of the authorities of the LPG and they gave him a piece of their mind and we had to give up our two fillies." In addition to cultivating the land and his full-time job at the LPG, Mr. Baumann also used his horses and carriage at weddings.

A second family, the Pflügers, cultivated a total of 7.5 hectares of garden plots of other members of the cooperative first with a horse and later with two dilapidated tractors. In addition to his regular work in the collective, Mr. Pflüger managed to raise some thirty pigs and twenty-five to thirty head of cattle, including milk cows. He even had his own milking machine. In addition to the feed he produced and the allowance he received when he delivered meat, he relied on collecting day-old bread placed in special containers next to apartment buildings in the neighboring city. Bread was subsidized and cheap, and so tons of bread were fed to animals. Like the Baumanns, the Pflügers had constant problems with the authorities. Mr. Pflüger was often called in front of the *Kreis* council. "There were always spies. When I threshed the grain and brought in ten loads of grain, the police would know how much I had threshed. I would tell them that I wasn't taking anything from anybody, that I had worked for it myself. Then they would be satisfied. I never was punished, but I did have to appear in front of them."[18] Thanks to his hard work, Mr. Pflüger was able not only to rebuild his own house, but to purchase and rebuild two houses for his children as well. A third family, two brothers and their mother who had access to a small, unspecified amount of land, fattened nine bullocks and kept nine to twelve sows and a few pigs to fatten for clients who preferred individually raised pigs.

One woman, Mrs. Feldmann, whose family history we summarized in chapter 2, refused to join an LPG and bought and resold eggs, an activity that precluded her from accumulating additional social security benefits, but permitted her to keep an eye on her senile mother-in-law. Her source for the eggs were the townsfolk who raised chickens and ducks on their garden plots.

To what extent did GDR-style individual farming prepare farmers for private production after 1989? From the foregoing discussion of marketing, farmers learned few marketing skills that they could have transferred to a market economy. Demand was assured from above and inputs could perhaps be negotiated through personal connections, but certainly not through market mechanisms. The lack of importance of market mechanisms under communism, even in those instances where there were private markets, is documented by Lampland (1995:344). Farmers in Hungary often preferred to sell their individual production, which became increasingly important in late communism, through the collectives or directly to the state, and were "far more interested in a guaranteed price than they were in the marginal differences in prices one could achieve selling at private market held in various sites across the country." Creed (1998:206) gives an additional reason why Eastern European, in this case Bulgarian, farmers shunned anonymous markets. He argues that in an economy of scarcity "to sell products on the market was to squander (the) alternate source of value" of engaging in more personalistic transactions that could "generate non-monetary as well as fiscal resources." Since selling a product could be regarded as a favor in and of itself, regardless of price, individuals preferred to sell to known individuals who could be relied on reciprocating such favors in the future.

However, the contrast between communist markets and modern markets in eastern Germany should not be overdrawn. As we shall see later, today, farmers often sell to monopolistic enterprises, and prices for commodities in the European Union are highly regulated through subsidies. Extensive experience with market mechanisms, while important, is therefore not as crucial as one might expect.

Whatever the importance of learning market mechanisms, individual production provided valuable experience in a broader range of agricultural activities. It is, therefore, not surprising that individuals who established independent farms after the *Wende* invariably had extensive experience with individual farming in the GDR.

The Family as Source of Identification

The opportunities for individual farming and the fact that the family continued to be the unit of consumption contributed to the fact that the family remained a major source of identification. The role of the family in agriculture was particularly pronounced among the individual farmers, who could never have managed to run sizable operations without the involvement of all members of the household. It should be noted, however, that such an intensive involvement of the family in individual agriculture was the exception rather than the rule in the GDR, a situation that contrasted with its prevalence in Poland (see, e.g., Nagengast 1991), its importance in Hungary (see, e.g., Lampland 1995; Vasary 1987; Szelényi 1988), and, to a lesser degree, its role even in the Soviet Union (Humphrey 1983); and with contract payment systems such as those practiced in Rumania, which gave individual farm families considerable independence in production (Kideckel 1993:109–116), or a similar system (*akord*) combined with the increasing importance of personal plots in Bulgaria (Creed 1998:94–108). It should also be noted that, with the exception of Poland, where most land was never collectivized, private production in all communist countries was highly integrated and dependent on socialized production.

In addition to its role in agriculture, the family in Germany was a haven from the invasiveness of the state, and family ties were important in mobilizing scarce resources (see, e.g., Dennis 1998).

The continued importance of the family did not, however, influence the power structure of the GDR, except perhaps in an indirect manner: it provided a remedy for the inflexibilities of a command economy, thereby both contributing to its perpetuation and sowing seeds that facilitated its eventual demise.

Leisure

"[When the collectives were established]," a former LPG manager recounts, "every farmer was able to take a vacation. That wasn't possible in a private farming enterprise. [In an LPG], if someone was on vacation, there were still all the others. Whether they worked in the cow barn or in the fields, every worker took a vacation. The length of vacation days was decided by each LPG. We had a minimum of eighteen days and additional vacations depending on the type of activity; for work that demanded particular responsibility up to four days could be added. Seniority was also a factor. If a worker had worked in the LPG for ten years, another day was added. As a result, a worker could get up to twenty-six workdays off.[19] The enterprises organized vacations in specific localities. We had vacation sites near the Baltic Sea, in the Forest of Thuringia, and in the Harz Mountains. In all these places we had bungalows or set up campers that we rented out for very little money to the members. In addition, we had a long-term lease on lodgings for six persons in the Carpathian Mountains in Poland, a well-known winter resort and two axle campers near the Baltic Sea located in camping grounds with toilet and washing facilities. In those camping grounds there were hundreds of other campers, both private and firm-owned. The private campers were mostly single-axle ones that could be towed by cars, whereas we had large ones that had be towed by a truck. At the end of the season, we would bring them home again. We rented two facilities in the Berlin Lake Plateau and one in the Forest of Thuringia as well as in Suhl. I think that there were eight to ten facilities all in all.

"Then we had a friendship contract with a firm near Nitra in Czechoslovakia. We had an exchange, whereby eight persons from here would travel to the High Tatra every year and eight persons from Slovakia would come here.[20] We would underwrite the expenses of the people from there and they, in turn, would act as hosts for our people. In that way, the arrangement worked without any problems. Every worker was allocated 300 marks for foreign travel, and each year we organized a trip to socialist foreign countries for one-fifth of the workers, that is, for 100 to 120 of the 600 workers. Each worker could choose which trip he or she wanted to take. That's how we got to know Carpathian Mountains, where we subsequently rented facilities permanently. We also went to Czechoslovakia, to Prague, and to the Bohemian Forest. At the end, we also traveled to the Soviet Union as far as Crimea, Moscow, Leningrad, Kiev, and Jerevan. All of this was paid by the LPG. The cost per person of such a trip would be some 1,600 marks. If a trip, say to Yalta on the Krimean Peninsula cost 1,200 marks, a person would have to save up his or her allowance for four years and would have to wait for five years to be able to go on the next trip. There was a record for each employee and one could get an advance if necessary. If a person left the firm with an outstanding amount, it would be deducted from the final pay." For such trips, the associated plant production and animal production LPGs would join forces and undertake the trips jointly. LPG managers stressed the feeling of community that resulted from such joint trips and celebrations. One of them concluded his description of such joint activities with the words, "Do you understand now why we couldn't just burden the people from animal production with the debts their LPG had incurred [during the communist years] and let them slide into bankruptcy?"

The exchanges with collectives in neighboring communist countries contributed to the establishment of wider friendship networks, although, as one consultant admitted, "Germans are a bit lazy when it comes to foreign languages. One always wants the others to learn German if they want something." The LPGs not only exchanged visits with collectives in localities in Czechoslovakia with which their towns had partner relationships but swapped spare parts and technical information. A butcher who worked in an LPG described these relationships as follows: "They came here and we celebrated festive occasions together and I prepared sausages especially for them and then we would barbecue the sausages and we would drink and they would go sight-seeing. They would remain for a few days. I was also there once with our manager. Three of us went. It was as though we formed a small state within the state. We three lived all alone in a large, perfectly equipped house. They would pick us up early in the morning and bring us back in the evening. Things like that. And when they came here, it was like that too." Two or three years before the *Wende*, mutual visits between the GDR and Poland and the Soviet Union also seemed to have increased.

The LPG Administration

The consolidation of smaller units into large industrial-like units was accompanied by the development of a large, complex, top-heavy administration in which individual bureaucrats followed (and/or circumvented) state directives, but where innovation was at a low premium. The management size and character was also related to the complex activities engaged in by the cooperatives, including many tasks outside the realm of strict production and by the continued role of private production of LPG members.

Former LPG managers often feel embarrassed when they are asked about the size of the LPGs' managerial staff. Among the factors that contributed to the bloating was the large size of the collective farms, the LPGs' dependence on the highly bureaucratized state apparatus for agricultural inputs, the controls placed on the LPGs by the state, the need to coordinate animal and plant production collectives, the need to engage in all sorts of ancillary activities to ensure timely access to material and service inputs, the lack or inadequacy of computing equipment, the requirement to keep a party secretary on the payroll, and the fact that the LPGs provided a host of services for their members and engaged in a number of activities in the villages or towns to which they belonged. For example, one 6,000-hectare LPG with a total workforce of 400 employed 10 managers and 20 to 30 staff workers who engaged in administration, accounting, and stock management.[21] These proportions were also reflected in wage allocation. Only 23% of the wages and other forms of remuneration were allocated directly to production (Gollnick et al. 1990:113). As one of the collective's former managers explained, "[The large size of the staff] was related to the fact that large stocks—stocks for the workshops—had to be administrated. Only one-third of the staff was engaged in actual plant production. The rest of the administration was employed in side production and ancillary activities. Every single brick used in construction had to be ordered, stored, taken out of storage, and accounted for. Like other enterprises in the GDR, agricultural enterprises were very complex because, in the planned economy, the division of labor did not function properly. If we wanted to produce, we had to see to it that we became somewhat independent of the uncertain conditions and the system interruptions." Our consultants agreed that, in spite of all the personnel, the complexity of the operation made it impossible for the firms to function very effectively.

The smallest production unit was the brigade, whose size could vary between ten and thirty members (Gollnick et al. 1990:40–42). It was led by a brigade leader who was absolved from undertaking physical labor. One or more brigades would form temporary or permanent work groups. The next organizational level was the section that coordinated the work of the brigades. In spite of the fact that workers tended to be highly specialized, an initial predilection for a product- and technology-oriented division of labor often ultimately had to cede to a more territorially based organization that saved on transportation costs, which was then promoted by the media.

The administrative personnel was also engaged in all sorts of cultural events within the collective for retired members and for the community at large. The following quote of an LPG administrator gives a sense of these functions: "I am ashamed to tell you [how many persons worked in administration.] But you have to remember that there were persons there who were responsible for all sorts of stuff, culture and things . . . that don't form part of agricultural production, but were assigned to the enterprise—senselessly, I would say today. We had a mess hall where cultural events, events for retired members, and events for the community took place. The groups were small, but there was somebody responsible for every little group." Gollnick et al. (1990:45) figure that, in a typical collective, 81out of 270 workers, or 30%, were not engaged in actual agricultural production. They (1990:46) outline the composition of a typical 5,000-hectare crop production cooperative without pastures or special crops as follows:

1)	Management (22 persons):	Head manager (1), production manager (1), security inspector, economist (1), office manager (1), section heads (4), brigade leaders (13)
2)	Production Preparation (5):	pesticide agronomist (1), fertilizer agronomist (1), dispatcher (1), feed economist (1), labor economist (1)
3)	Administration (10):	Principal bookkeeper (1), bookkeepers (8), secretary (1)
Total (37) (13.7 full-time equivalents), Culture and Social section (15)		
4)	Workers	Tractor drivers and hand laborers (189), construction brigade (14 workers), repair shops (15:5 for each of the three territories)

The fact that the state had little interest in promoting the autonomy of firms contributed to the inefficiency of the operation. One former LPG manager explained: "We didn't have access to proper office equipment. While we did have a computer, it was only an 8–bit-computer with which one couldn't carry out a proper expense analysis. Everything had to be done by hand, more or less. And, even if we had owned a computer, we would not have been allowed to [engage in such calculations] because the accounting was done at a central computing location so that the state could always see what our finances looked like and remain informed."

Administrators felt more constrained in their ability to make decisions than they do now. For example, a former manager of an LPG tractor workshop who now acts as a submanager for a tractor distributor felt that he had much more freedom to make decisions in his present job than he did in his position in the LPG. "Before, we were relieved from almost all decision-making and that is no longer the case now. Then, whatever the decision, it was always wrong. . . . I had to defend every decision I made. I would conceive a project proposal and would then present it, saying what I required, whereupon it would be discussed with this or that person. Then, a new idea would be proposed. Someone would say that we didn't need [what I had proposed], but we needed something else. That would be called a 'mechanization concept.' That concept would then be elaborated on and discussed repeatedly. After that, it had to be approved and money requested to carry it out. Ultimately, we would receive only part of the requested money, which meant that we would have to decide what we could do without and present a new proposal. But then the machines would not be available and we would suddenly have money left over and would have to try to do something else. I would say that that (the ponderous nature of the allocative process) was the greatest impediment."

The incentives for an LPG manager to innovate were low. Salaries for top LPG managers were not more than twice the wage of a tractor driver. Official recognition went to those who kowtowed to the party, not to innovators. The same consultant quoted above put it this way: "I didn't get anything out of it except work and headaches. My sons told me once, 'Father, in another type of society we would have a millionaire as a father; but, as things are, we have a father whom we have to beg to buy us a bicycle.'" For his pains, an innovator could only expect reprimands. "I was subject to all sorts of punishment," the same consultant continued. "I never received any recognition from the state. I have no national prize. I have nothing. On the other hand, it wasn't the case either, as many think, that if one vented one's frustration and said that things could not continue like they were, someone was already standing behind you to lock you up. That is nonsense." He claimed that what had motivated him to innovate in spite of the lack of financial incentives was the desire for self-realization, the feeling that he could achieve something on his own.

Another former LPG manager gave a more nuanced description of the relationship between the LPGs and the Communist Party. "The situation was like this: we couldn't do anything against the party, but the party could not oppose the will of the cooperative farmer either. The party had to approve a candidate for office, but, in turn, that person could not be deposed without the consent of the cooperative unless he was found guilty of a crime. It was similar to the [UN] Security Council, where members enjoy complementary vetoes. So there was some democracy here in these firms. . . . Also, if we didn't agree with directives we just worked to rule. Then we said, 'Fine, we will do it,' acted as though we were complying, but didn't actually comply with the directive."

The LPG administration was supervised by a board of trustees (*Revisionskomission*) whose members were, in part, members of the collective and, in part, outsiders. While the composition of the board could lead to heated discussions, particularly about the amounts to be distributed and the amounts to be placed on the reserve account, day-by-day decisions were usually made by those who were versed in agriculture to the extent that they were not dictated from above. As one former LPG manager said, "We could not decide to plant more potatoes [even though that might have been to our advantage]. Neither could we choose to keep only 550 cows instead of the 700 dictated to us, although keeping a smaller number would have resulted in a production of 4,000 liters per cow instead of only 3,000. We produced 3,000 liters per cow for years and everybody thought we were too stupid to produce [more]. Such things happened all the time."

Who became the manager of an LPG was, in theory, determined by its members. Indeed, as we have seen in chapter 2, in at least one collective, the manager insisted on holding elections by secret ballot. However, while in some LPGs nominations for leadership positions could come from the LPG brigades and the membership sometimes could force the party to accept an inside candidate, the party had a determining voice in the matter, and at least one former LPG worker claimed that the party alone made all the managerial appointments. "We were not asked to give our opinion. They would tell us all sorts of stories, so that, at the end, we would be persuaded and either gave our vote or not. The names were all predetermined. No one admitted it, but that's how it was. Whether one raised one's hand or not, they would say that the vote had been unanimous and that was that. Most of [the workers] just went to vote because they served sausage and coffee. . . . After all, it was already predetermined who would become a master craftsman and who would be appointed, who would do this and who would to that, or who would go to school." Even the former LPG manager who insisted on holding elections by secret ballot conceded that "in most instances the *Vorstand* (executive committee) was elected after the party had suggested candidates who were then voted in in an open vote. When 240 persons sit in the hall and 'the man from the *Kreis* council,' as he was called, presented the candidates as the best, no one would dare to vote against them. Abstention didn't exist. Before the *Wende* there was no abstention in the GDR; one had to vote for or against something. Then, the vote would proceed. Say, there were 200 persons in the hall, they would call out, 'who is in favor?' and, perhaps 130 would raise their hands. Seventy would not raise their hand, either out of persuasion or for some other reason. Then they would ask, 'who is against?' and perhaps 5 would raise their hand. So 195 were in favor."

Membership in the Communist Party was, in most instances, crucial to be considered for top positions in the LPG. The state continued, however, to maintain the figment of a multiparty system by allocating a fixed share of these positions to members of the bloc parties and, on occasion, even to individuals who did not belong to any party. The process was described to us by a former LPG manager as "a formula that was not written down anywhere, but existed nonetheless. For example, if an LPG manager retired and he was a member of the Farmer's Party, they would try to find another member of that party to replace him. If that was not possible in that particular LPG, one could place such a person in another LPG. The proportion was more or less one person who did not belong to any party to two persons from the bloc parties and ten to twelve from the SED. But one can't say that the latter were necessarily the lesser qualified individuals. Good and ambitious persons often recognized that they could only get ahead if they joined the party. So they would join. It was totally insignificant whether a manager belonged to the party or not. He had to follow certain guidelines anyway. Someone who did not belong to the party couldn't act any differently from one who did."

The credentials of the heads of the agricultural cooperatives we interviewed, who had all either managed the LPG out of which the cooperative had been formed or had acted in some other managerial capacity within it, were indeed impressive, irrespective of their party affiliation. One former LPG head and a former middle-level manager had been professors at agricultural universities. One cooperative head who formerly worked in the same LPG as a specialist in agricultural economics attended an agricultural school for three years, where he specialized in seed raising, and then studied agriculture at a university, where he received a degree in agriculture at the end of a five-year period during which he alternated one semester of full-time study with one semester combining work as a consultant in the sponsoring LPG with taking correspondence courses. Thereupon, he took correspondence courses to complete a second degree in economics. Two other former LPG managers had a similar background, but without a second university degree. And a sixth former LPG manager studied tropical agriculture at the university level and, for many years, worked in development projects in the Third World.

Superior credentials and even party membership did not protect our consultants from the heavy-handed interference of the state in their lives. Pressure was brought on those with relatives in the West to cut off all ties to their kinsmen. Visits to the West were denied, except during the last year before the *Wende*, when the party began to loosen its reins ever so slightly. The two former professors both lost their university positions because they were unwilling to testify against a colleague or refused to sign a document supporting military intervention in Czechoslovakia after the Prague Spring. As the example in chapter 2 showed, additional forms of harassment included a raid on an LPG in the middle of the night to check if the management observed all regulations, a raid that coincided with a request to travel to the West.

LPG managers had to certify the political reliability of young apprentices whom the state was recruiting to patrol the border with

West Germany. And managers who were party members had to attend endless meetings, which included a "day of party activism" at sowing time and another during the medicinal plant and beet harvest. The annual cycle culminated in a weeklong retreat where each LPG was scrutinized for of its fulfillment of the five-year plan and was subjected to "exposure to red light" (*Rotlichtbestrahlung*, which also means X ray), which entailed a day of heavy communist propaganda. This event was recalled with mixed feelings by a former party member and LPG manager: "We had to arrive on Monday noon for the one-week retreat, but those four or five days were really rather nice. The worst day was Monday. The secretary for agriculture of the *Kreis* would speak about the role of agriculture and the prognosis for the coming year. That prognosis was never correct, but he had his say and then he returned home, allowing us farmers to be among ourselves. One or two professors would come to speak on different themes. We would have long discussions and play Scat (a card game) in the evenings. We also went for a long hike followed by a meal of trout. On Thursday evening, we had a farewell party, so that there was some event every evening, and on Friday noon we either drove home or were driven home. Except for Monday, it was really a nice week. . . . The comrades in the *Kreis* went to great lengths to make it a pleasant week. It was like a ritual. The first person to speak was someone who had no idea about agriculture but then we were edified."

The time spent with political activities varied considerably depending on a manager's ideological commitment and willingness to follow the directives from above. Party members who enjoyed particular trust in the party were forced to take part in a larger number of events and had to engage in frequent travel to meetings. In contrast, those with relatives in the West were exempt from such activism.

Farm Manager Identities

The identities of the managers of collective farms, our principal informants for the GDR period, corresponded with their special situation within the GDR economy. In many ways, they were no different from the managers of public enterprises. They were part of a chain of command that determined priorities in production and the very form of agricultural enterprises. Even though the collectives had some say in electing officials, they could be imposed from the outside. Often they did not even own land in the collective. They had little leeway in dividing profits, for surpluses were largely syphoned off by the community and the state. And yet their position of power could be used to divert resources to fulfill their personal needs. For example, while the collectives' construction crews regularly built houses for the members, the managers' own homes could receive priority.

And yet, perhaps more than in state-owned and -operated enterprises, the managers of collectives did put their own imprint on the enterprises they directed and on the farms' relationship to the wider community. They viewed these actions as evidence of their independence from the dictates of the system. First, compensation—which, at least during the 1980s, was competitive with and sometimes even significantly exceeded compensation in industry—was still to some extent related to productivity. They could attempt to circumvent, often successfully, the most harmful directives coming from above. Shrewd accounting could increase the distributions to members to some extent. The managers of collectives we spoke to thus regarded themselves as more than cogs in an overarching bureaucracy.

Some of our informants who had headed collectives saw themselves as having been rebels in the past in more unorthodox ways. Two former professors had become managers of particularly run-down collectives after they had fallen into disgrace with the communist regime. One of these, who had lost his position at a university because he did not give wholehearted support to the GDR regime's move to discredit Dubcek, said, "I never belonged to any party. I also didn't hide the fact that I equated communism with religion. There are so many similarities with religions even in the way it manifests itself. Think about those large demonstrations. They were the processions. Pious hymns were sung to accompany large portraits of the prophets Marx and Engels. They too had flowing beards, just as in the Bible. With his bald pate, the founding father, Lenin, looks like Saint Peter. The only thing missing was the halo. If [communism] had lasted another 200 years, the halo might have been added too. . . . Think about [the similarity between] the cardinals and the politburo, about all those old men at the top, or about the belief in miracles. Although we were too enlightened to believe in the latter, we did have the huge competitive drives, the economic miracles. . . . Even the small details [reminded me of religion]. The Catholic Church continues to have confession. Well, when someone committed some error, even in his private life, in the family or in his marriage, the SED (Communist Party) forced him to engage in self-critical reflection in party assemblies."

This consultant had relatives in the West, which also raised a red flag for the ubiquitous secret service. But when asked whether the fact that he had relatives in the West made his life difficult, he answered, "So schwierig war es nicht. Ich habe damals gesagt: 'Ist der Ruf erst ruiniert, lebt es sich ganz ungeniert' (It wasn't so difficult. I used to say, 'Once one's reputation is ruined, one can live quite shamelessly')."

Even the only die-hard Communist among our farmer informants—who, although he had become a successful capitalist entrepreneur after the *Wende*, praised many aspects of the communist system—saw himself today as having been a rebel in the past. He considered his success as the originator of a processing plant to have been the result of his extraordinary efforts at manipulating connections and his ability to establish a unique niche where he had a degree of autonomy from state dictates.

The managers also had considerable impact on surrounding towns and villages. Like other large enterprises, agricultural collectives constituted communities that provided for the needs of their members both inside and outside the work situation. As De Soto and Panzig (1995:186) explain,

> In contrast to West German villages, which exhibit a loosely structured community social organization (i.e., school, church, mayor's office, voluntary associations such as *Vereine* or clubs, etc.), and which transmit a sense of community identity through local rituals and remembered traditions, in rural East Germany economic and cultural identity was closely connected to and predominantly constructed by one's collective. Over the years, for many women, the collectives became a multi-functional social and cultural center. (We would add, that this was true for men too.)

First, as in all enterprises, whether run as state enterprises or as collectives, the role of the managers was one of creators or, at the very least, upholders of jobs. They were forced to accept the joining of less profitable enterprises with profitable ones without recourse to firing. Since surpluses were largely taxed away and could not be distributed to workers, there was little incentive to reduce the labor force. Instead, they created facilities ranging from repair shops to vacation facilities that made them less dependent on the central allocation process of services provided by outside agencies (see Kornai 1992:223; Verdery 1996). The managers' ability to contribute to community projects and their positions on town councils did give them a degree of personal clout in the region.

The image that former managers of collectives have of their role in the past is thus as shrewd mediators between the dictates of the party and the needs of their workers and those of the wider community. While beholden to the hegemonic power structure, they regard collectives as having been very different from state enterprises. Although many of them were members of the Communist Party, the SED, and a much smaller number of the bloc parties (whose platforms were never allowed to stray far from that of the SED), they were able to relinquish the more propagandistic aspects of party rule to the special representatives of the party who were placed into every larger enterprise. As a result, the managers could distance themselves from the role of upholders of the state ideology and, at least in retrospect, construct themselves as spokespersons for the members in the workplace and the community.

PART II

The Post-Communist Period

INTRODUCTION

The Aftermath of Reunification

In Part I we explored the tension between the hegemonic state enterprise model and the collective model prevalent in agriculture. We also analyzed the impact of this situation on the ways in which managers of collectives constructed their identities.

The demise of the GDR led to the dissolution of state-managed farms and the questioning of the collective model. At the same time, precommunist actors reemerged and new actors came from the west. In Part II, we will discuss this process, beginning with an analysis of the immediate aftermath of the fall of the Berlin Wall and the subsequent reunification of the two Germanies, followed by the struggle of collectives to reinvent themselves as capitalist enterprises, the ascendance of individual farmers from both eastern and western origins, and the clash among these various actors. We shall argue that this clash has forced the actors to define their identities in opposition to one another as well as in relation to outside forces, particularly the *Land*, the nation, and the European Union. But, just as in the GDR the various forms of agricultural production, while entailing incommensurable differences, also complemented one another, the divergent forms of agricultural enterprises that sprung up after the *Wende* also serve complementary needs. The joining of the two Germanies must therefore be regarded in ways more nuanced than the model implied by a clash between two irreconcilable systems. As we shall see, all actors have had to adapt to the dictates of international economic competition in general and the vagaries of the Common Agricultural Policy of the European Union in particular. Their competition also entailed emulation, narrowing the gap between them, while not eliminating it entirely. Also, on occasion, actors belonging to one category, assume the identity of another. Finally, the actors also sometimes cooperate.

Analyses based on Cartesian dualities, such as Tönnies's contrast between *Gemeinschaft* and *Gesellschaft*, fail to provide an adequate rendering of transformations like those in the former GDR for an additional reason. Both communism and capitalism have heterogeneous origins and some of the strains in their development overlap. Western cooperatives and eastern collectives both evolved out of the same nineteenth-century forms. The different actors draw from this common past. Furthermore, they also all draw from those aspects of their shared precommunist past that had been discontinued or partially submerged in the East during the period of separation. Sometimes, the actors even draw from aspects of this past that were obliterated or held in abeyance in *both* East and West. As in the case of GDR agriculture, agriculture after reunification exhibits contradictions, paradoxes, and overlapping structures that require intricate models of description and interpretation.

As we have seen, the state of agriculture in the GDR at the time of the *Wende* was better than that of other sectors of the economy. We would agree with one LPG manager's qualified assessment that agriculture was not in imminent danger of collapse and could have survived for some time longer under the political and economic conditions that prevailed just before the fall of the Berlin Wall, even without price supports. "We could have continued to have operated as an LPG for another five years. If we had engaged in barter or something else, it would still have worked out." Indeed, there were few indications that agricultural workers joined the exodus to the West in 1989 in any significant numbers.

The relatively good condition of agriculture did not mean that farmers were indifferent to the events of the fall of 1989. One LPG manager recalled that tense time, during which weekly demonstrations took place in Leipzig and other major cities in the GDR, in these words: "My siblings and parents live in Leipzig. We phoned each other every evening and so we were able to keep abreast of events there. My brother works as a physician in Leipzig and my sister is a physiotherapist, and they knew what was going on. They also knew just how great the danger was. I remember my mother calling one evening and saying, 'Now it's going to happen.' Everybody was afraid. My mother had heard from my sister that blood was being shipped to the hospitals in Leipzig. . . . The physicians were expecting the worst. In such situations an almost traditional authoritarian standard prevailed in the GDR. Any self-respecting hospital made certain that it had an adequate blood supply. That was a matter of honor. . . . What is remarkable is that not a single drop of blood flowed. They confronted one another face to face, and the machine guns also stood on the houses in Leipzig and Berlin. But no tanks rolled into the city. Nothing happened."

We did not ask farmers to reexperience the opening of the Berlin Wall, and unlike some of artisans—the dramatic description by a baker's wife of a sea of headlights moving toward Berlin in a slow but highly organized stream in both lanes of major highways during the night of November 9, 1989, stands out in our minds—neither did they volunteer to relive that momentous event. Particularly in the case of beleaguered managers of cooperatives, it seemed indelicate to probe too deeply. We felt that regardless of their sentiments at the time, the subsequent economic turmoil that threatened the very survival of agriculture in the new *Länder* in general and the continuity of agricultural collectives in any recognizable form in particular would be a subject of much greater concern to them. Although they recalled the horrors of unemployment, the market crash, the useless feasibility plans, the legal maze, and the ever present possibility of dissolution, the leitmotiv of their tales was their sense that they could survive the onslaught of a different system by a judicious bricolage. The proof was that they had indeed made it through the first few years (our informants were, of course, the successful managers). One manager of a cooperative did, however, comment about his involvement in the political process with the opening to the West. After explaining to us how he had kept abreast of production processes in the West should he ever have the opportunity to export to the West directly and to obtain hard currencies directly, he continued, "One of our members always gave me long speeches about the beautiful tractors they had in West Germany. For the fun of it, he would go and plow without pay for a farmer for half a day just so that he could drive one of those nice tractors. 'Oh, if we only had those,' he would say. I would answer, 'When we have those, we will no longer exist.' Now we have the tractors. The man is still with us too. And we probably have more money than the farmers who owned the tractors that he drove at that time. So he told me, 'Can you understand such a thing, chief? Now we have the tractors and we are still alive.' Well, what happened? The GDR people were stupid. The masses are always stupid. They thought, 'We have work, we have lodgings, and now we will get Western money. Then we will prosper. What could possibly go wrong?' And Mr. Kohl promised everything.

"In February 1990, I was on television preparing for the election campaign—I belonged to the Farmers' Party—and I said, 'Let's form a German confederation immediately, so that changes can't be rolled back.' I am a traditional German who felt like throwing up when I thought about the regime. It was alright with me if Weizäcker became our president. But then we would need the free convertibility of eastmarks into westmarks, open the borders, free commerce, and change with a transition period of five to eight years in the state and the entire confederation. If one did it too quickly, I said in just these words, I feared that there would be 2 million unemployed workers. I knew what was coming. But things became much worse than I had feared. In reality there are 3.5 to 4 million unemployed in this region when you add the persons who retired early, ABM (workfare program) workers, those who receive transition to retirement payments, and all those things. In principle, the economic base of the population here has been broken into pieces. In some ways, it's like it was in 1945 when the Russians came." The shock of reunification has formed this cooperative manager's sense of himself and others like him as the next section reveals most graphically.

The Crash in the Markets: Panic Purchases

The opening of borders brought an immediate flow of goods sold at exchange rates of 2 to 3 eastmarks to the deutsche mark. The availability of these goods (to the extent that they could be paid for with scarce deutsche marks), in turn required a reevaluation of what things were worth. As a former manager of a tractor pool of a state farm and present assistant manager of a tractor dealership explained, "At the beginning everything was new. We were thrown into icy cold water. Traditional understandings of prices no longer held. For example, before, an apple might cost a fraction of what a banana cost, when bananas were available at all. Now, it's the other way around. In the same way, technological goods were very expensive, because electronics were prohibitive." And yet, by mid-1990, at the time of monetary union, many easterners still did not understand the new measures of value. Apprehensive that the rate of conversion of the eastmark to the deutsche mark would be unfavorable, people stood in line in front of the stores in order to convert their money into anything of value. According to the same informant, "[At the time] people made purchases out of sheer panic. They bought bed linen and towels by the closet full. They bought video-recorders for 6,000 or 7,000 marks in the belief that an electronic gadget represented an investment. They stood in line for entire nights, because they wanted to invest their money somehow. When there were no video-recorders left, they would simply buy a washing machine."

The public's reaction was understandable. As we have seen in chapter 4, in the GDR everything except for the most basic necessities was scarce and therefore valuable. "I actually know several persons who have coal-fired boilers standing in their homes," our informant continued. " Now no one wants them any longer.[1] But that was an item of value. After fifteen years such boilers would break down. . . . So, when there was something to be had one would buy it. The money was there, but the goods were not. . . . Where would we have been able to obtain the information [about such matters]? We never were able to go out [and see for ourselves]. Many persons lost their entire savings in this manner." On the other hand, it would have been difficult to invest in something that would have kept its value. "One couldn't buy land. That was the problem I had. I have a house lot of 500 square meters and behind it is a garden of another 700 square meters [that I have rented]. I was able to buy the house lot, but the legality of the transaction is now being contested. So one doesn't know what will happen with the land. They say that they won't be able to take away anybody's house because we built them ourselves, but what about the land? Normally, when my house stands on land that doesn't belong to me, then the house doesn't belong to me either. Thousands of people in the east are affected by this issue."

New Markets and Pricing Structures

Once eastern Germans had the choice, no one wanted to buy local goods or even local produce. As one consultant put it, "Right after the *Wende* one couldn't get a mark for a Trabant or a Wartburg car. People let them stand and would go and buy a fifteen-year-old car for 10,000 from some wild used car dealer just to own a car from the west. That car would break down on the way home. Then, at one point or another the money would be gone and they would go back to driving a Trabant or Wartburg after all. Their value went back up and now one can get 500 marks for a Trabant or 1,000 for a Wartburg. People are beginning to recognize that those cars are worth something. They don't have the quality and so forth, but some value is there. Things have become more normal again." In the same way, farming cooperatives at first refused to buy any tractors from the east. So western dealers were soon able to establish themselves. One dealer delivered ten tractors and machines even before monetary union and leased large numbers in the first years after the *Wende*. The large tractor factory in Schönebeck stopped producing tractors, and the large Russian tractors that were used to lug missiles, while still in demand, did not have the features of their far more expensive western counterparts. Indeed, accustomed as they were to fixed prices, eastern farmers often were talked into paying higher prices for western technology than the dealers charged their western customers. Interestingly, at the same time, there was a booming market for used eastern German tractors and other farm equipment in western Germany. The small-scale farmers there seemed less concerned with breakdowns than the large-scale eastern farmers. For the same reason, they were more likely to purchase western tractors after they had been leased to eastern farmers. As a result of the unloading of eastern technology, workshops of cooperative farms found themselves with large inventories of spare parts they had carefully hoarded before the *Wende* to preempt shortages. One state farm sold them at bargain basement prices to farmers who still owned eastern equipment.

In the beginning, farmers faced major adjustments when their products were no longer in demand, access to western markets was closed for many of their products, especially for fresh produce, and prices crashed for all commodities, but especially those that were formerly heavily subsidized. Between 1989–90 and 1992–93 crop prices fell between 38% and 76%, and wheat, the region's most important crop, fell 49.7%. Animal product prices fell between 65% and 77.4% (Thiele 1998:27). For instance, the price of wool no longer even covered shearing, while mutton, never popular in eastern Germany, could not be marketed without special connections with clients in the west. One cooperative manager looked back at the decision to get rid of the farm's large herd of sheep with sadness. "We used to have 1,500 merino sheep. We got rid of all of them, because they are not profitable. No one wanted the wool, or the meat, or at prices so low that they could only have been a sideline. In addition, we had other problems with sheep. The sheep stalls are in the middle of the village, so we always came into conflict with community ordinances. The sheep have to be taken out every day and, of course, dirty the streets. [If we wanted to keep raising sheep], we would have had to build new barns. But at current prices, this would have been even less profitable. We had to get rid of them, even though I regret this very much, because here sheep raising does not only have an economic dimension. We have the river embankments of the Saale here. Who is going to take care of the grass that grows there? The sheep used to take care of that. The same is true for hill slopes that are almost impossible to mow." As we shall see, the few shepherds still around are eking out a precarious existence or are engaging in sheep herding only as a secondary activity.

At the same time, major changes had to be made in cattle raising. The market for many milk and meat products fell precipitously as novel, better-packaged products came from the west and the traditional outlets, the *Konsumgenossenschaften*, collapsed. Between 1989 and 1992 gross production was reduced by 36% (Schmidt 1992:14). At the same time, for reasons to be explored later, herd efficiency improved dramatically but the total output could not be increased because milk production was subject to quotas. Finally, since workers could not be dismissed from one day to the next, the sale of livestock was used to pay wages. The combination of these factors and, in 1992, one of the worst droughts in decades, led to a massive sell-off of cattle and pigs, sometimes at extremely low prices. Fortunately, the glut that the resulting flooding of the market could have occasioned was mitigated by an accord the German government reached with the Soviet Union to purchase large quantities of meat, grain, butter, eggs, tobacco, and cigarettes.

In the case of plant production, eastern German farmers had difficulties finding markets for their fruit and vegetables. Not only was it difficult to compete with the Dutch, with their humid climate and long experience growing vegetables in small spaces, but the easterners had to contend with negative perceptions of the quality of their produce as well as competition from low-wage producing countries and western grocery chains committed to western suppliers. Some of the producers took advantage of state payments to rip out orchards. One cooperative manager remembered, "They paid a premium to cut down the [apple] trees. We had profit margins of 50 fennig [per kilogram]. In addition, we had the storage costs, leaving only 9 pfennig, for which one can't produce a kilogram of fruit. We had no sales. Everybody wanted apples from Holland, the Tirol, or the Lake of Constance. They didn't want to eat those 'poisoned' apples any longer. But then, when we talked to our colleagues from the Lake of Constance, we found out that we were, according to them, not using *enough* pesticides. We sprayed ten to twelve times per year, while they sprayed eighteen to twenty times." Another cooperative manager was equally frustrated: "After the *Wende*, I thought that the cooperatives would be shown a meaningful way to compete. That was not done. Instead all sorts of things started to happen right away. They paid premiums for each tree that was torn out. So orchards were cut down and one D-mark was paid for each tree. Everything that constituted a competitive threat for the FRG was immediately destroyed. 'Away with it, we don't need that any longer.' We were not the only ones to be affected; thousands of hectares of trees were ripped out. They paid a premium for that! They apparently didn't fit into the picture. Think about the Werder/Potsdam park. Those who have read the history of Potsdam know that already at the turn of the century people would go to Werder when the cherry trees bloomed. Now they have uprooted them. It's all gone."

When the same cooperative planted carrots and peas, no one wanted them. The factory that had processed vegetables before the *Wende* was closed. "At the time of the *Wende* we had a terrible experience. We had planted 150 hectares of canning peas and 70 hectares of carrots. The harvest was ready at the time of the *Wende* but the factory did not survive and there were no other processors in the area. Agricultural production was quite good in the GDR but processing was miserable. The factories were obsolete, already a hundred years old. We were stuck with the entire harvest of peas and carrots. That was a big loss. We tried desperately to get rid of them, even in the old *Bundesländer*. But they were not equipped to deal with such enormous quantities. We were able to sell 20% of the peas by driving around in the entire region. It didn't work out at all with the carrots. They rotted in the fields. The hares sure had a good winter that year!" Another cooperative that had exported onions of a certain small size found that the demand for these onions had evaporated after the *Wende*. The same cooperative was able to sell its last 1,000 tons of potatoes only with great difficulty. "We drove around the cities with trucks, ringing a large bell and carrying banners, but in the evening the trucks were still half full. No one wanted to eat GDR-products." Not surprisingly, in the year following the *Wende* (1991), farmers reduced the amount of land planted in potatoes by 58.5% in the new *Bundesländer* (while there was an increase by 4.3% in the old *Bundesländer*) (Top Agrar Spezial, October 14, 1991). The demand for hothouse vegetables vanished from one month to the next too. Thus one cooperative lost its market for cucumbers and tomatoes entirely to imports from the west.

Producers of specialty crops had similar difficulties in finding new markets. Many of the herbs, for which the region was known before the *Wende*, which had been the basis for the more than average prosperity of some of the farms and for the relative autonomy from official meddling, could be produced more cheaply elsewhere. The fish-processing industry, once the main client of the herb oil factory, the pride and joy of one of our former LPG manager informants, had collapsed; the future of the cosmetic factory in Berlin, which had been another major client, was still uncertain; and new markets were difficult to penetrate with this relatively exotic product. Also, the industries that had been taken over by western firms relied on the western suppliers connected with these firms rather than continuing their ties with their traditional eastern sources.

Even when there was a demand for their products, the farm managers often did not get the best prices for them. Just as they had difficulties bargaining when they purchased western farming equipment, they had to learn not to accept the first offer they received from dealers interested in their products.

Finally, farmers were sometimes duped by dishonest foreign investors. Before they realized what was happening, their animals or their money would be gone. Sometimes livestock was even stolen outright. "There were instances where a cattle truck appeared at night and loaded all the livestock," exclaimed one informant. "We were spared, but it could have happened to us too because two of our barns are located away from the beaten track. No one would have paid attention to an extra truck coming to load animals at night; for we often loaded animals at night ourselves."

Government Supports and Useless Feasibility Plans

The former LPGs received substantial government assistance to tide them over the transition to capitalism in the form of compensation for the sudden decrease in prices, support for farmers in agriculturally marginal areas, and ensuring of liquidity. The first type of support amounted to 1,100 deutsche marks in the second half of 1990 and 1,400 for 1991 for each full-time worker.

In order to qualify for financial support, an LPG in transition had to prove that its assets per worker did not surpass a certain limit and that steps were being taken to adapt to market conditions by providing a production plan (*Konzept*) for the agricultural year. LPGs with more serious liquidity problems also had to provide detailed accounting and a restructuring plan. A plethora of advisors with very uneven competence flocked from the west to offer their services. The scene was described by a cooperative manager: "In order to tap these bridging funds we had to present a plan to the department of agriculture of the *Land*. A plan was needed just to be recognized as an enterprise. If it looked pretty [with lots of diagrams], they accepted it. People emerged out of the woodwork to create such plans who had no idea of what they were talking about. We would provide them with numbers, as best we could, and they would then create a plan for the enterprise. I shudder, when I look at the first one that was designed for us. It was beautiful. A person from the neighboring farm told us that someone had made a good-looking plan for them and we thought he could do that for us as well. So the man called on the Friday before Pentecost and said that he would come on Sunday at 1 P.M. He stayed on Pentecost Sunday and Monday. On Wednesday he left after he had received his 13,000 deutsche marks, or whatever it was, and we had a beautiful, multicolored plan. In the mornings until noon we would have to answer all sorts of questions and in the afternoon he would stick everything into his computer. I am sure that he had some kind of software/hardware and the computer would spout out everything beautifully, with diagrams. He had it all bound and we presented it to the authorities. Everybody was thrilled with it. One could do nothing with it, but it looked good. The next year we made a new one and last year we made a proper one. It is a seven-page document made by Finanzagrar. With that one, we can now actually do something. One of the reasons I can get more out of these plans now is that I now understand the terminology. Many of the terms were entirely new to us then. For example, we had no idea what the term *Deckungsbetrag* meant. It means [gross income minus variable expenditures]. We knew what profit meant, but *Deckung?* People would come and say, 'You have to reach such and such a *Deckungsbetrag* per hectare of wheat and barley.' Now we know what it means. [But then] we would have to ask them, 'This is the income in deutsche marks, tell us what we have to subtract in order to get to our figures,' and then we would know what they meant."

The Maze of Legal Forms of Ownership

As unsettling as the upheaval touched off by economic unification with a country with economically and politically powerful commercial networks was, the issue of the future management and ownership of agriculture was of much more lasting significance. The use of the land per se in some, usually intensive, form or another was assured for all but the most marginal land, mostly located in the north, although set-aside premiums were paid to farmers who fallowed land or converted it into forests. Rather, what was at stake was whether the LPGs would be allowed to continue under another legal status or whether individuals would take control of the land. At issue, as well, was the optimal size for the restructured firms and the tasks they would undertake.

Concrete moves toward privatization came as early as the first months of 1990, when the Modrow government that succeeded Honecker and Krenz set up the *Treuhand-Gesellschaft* (usually referred to simply as the *Treuhand*), an agency that took over the administration of all government assets and sought to privatize viable firms while liquidating the rest.

Agriculture constituted a special case. Land ownership and land use were in consonance only in the case of state farms. These enterprises, whose structure resembled that of state-owned industrial complexes, cultivated former demesnes of the *Länder* and some of the land expropriated from large private landowners in the late 1940s. Only there was the state free to impose its will on an enterprise. Indeed, such enterprises were systematically liquidated. In contrast, agricultural collectives, the LPGs, could not be placed into the same category. The land they cultivated also included land that had reverted to or had been expropriated by the state, or had belonged to the *Land* in precommunist times. But they were structurally similar to the legally recognized cooperatives in the west because much of the land they cultivated was technically still owned by members whose income continued to depend, to a degree, on the net profit of the collective. The ambiguous status of the collectives led to an equally equivocal response on the part of the post-*Wende* administration. While it was legally difficult to justify mandating liquidation when two-thirds of the members of a collective opted for reorganization in a recognized corporate form, the new government officials often felt hostile toward the choice of a cooperative form of farming. On the one hand, the clear legitimation of agricultural cooperatives by the *Treuhand* during the de Mazière government was reaffirmed in the 1991 Law of Agricultural Adjustment (*Landwirtschaftsanpassungsgesetz*) which states that, "All forms of ownership and production, the family farms and cooperatives voluntarily formed by the farmers, as well as other agricultural enterprises are given equal competitive opportunities" (quoted in Luft 1997:16). On the other hand, the cooperatives' continued existence in any recognizable permutation of their erstwhile form was and continues to be questioned, although, perhaps because of the fact that a social democratic government came into power in 1998, cooperatives feel more accepted now. The resistance to agricultural production cooperatives notwithstanding, cooperatives in the former GDR have fared well in comparison to those in some other eastern European contexts. Thus cooperatives in Bulgaria were summarily dissolved when the Union of Democratic Forces Party gained power in 1991 (Creed 1998:219) and rejoining cooperatives was made difficult (224). Interestingly though, the losses suffered by the community studied by Creed in the process of liquidating its cooperative "strengthened cooperative sentiment among the villagers—the very opposite of its intended effect," and a new cooperative was formed in 1992 (239). By 1994, the vast majority of farmers had rejoined a cooperative (250). Similarly, decollectivization was instituted by the Hungarian government in the early 1990s (Lampland 1995:351). In contrast, in Rumania, farmers had the choice to remain in cooperatives or to privatize (Kideckel 1993:218–222).

As we have seen earlier, land had different legal status depending on whether it was brought into an LPG by smallholders who had inherited it or had obtained it through the *Bodenreform* of 1946, or whether it had been expropriated from large landowners and not redistributed, or had reverted to the state or the commune because the owners had ceased to work in agriculture. While the form of ownership—private or public—or the history of ownership made very little practical difference in GDR times once the process of collectivization had run its course, this situation changed radically with unification. Inclusion in the official registry as a private owner became meaningful once more. Land so registered could suddenly be disposed of by sale or leasing arrangements. Since the possession of a title to land did not entail much of an advantage in GDR times, owners of *Bodenreform* land were sometimes not overly concerned if, for some reason, their land was not registered (e.g., when it was transmitted to an heir). Such omissions are presently causing hardship for the affected individuals. Conversely, some individuals who still figure in the registry, even though they left the cooperative and should, theoretically, have lost their claim to *Bodenreform* land, have received an unexpected windfall. In 1994, no final determination had been made by the courts regarding unclear titles, but one cooperative manager expected that those who could make reasonable claims of ownership even to unrecorded land would be able to have their land entered into the registry after the fact.

The post-*Wende* status of the land that LPGs had received for cultivation during the GDR from the state was less clear. The *Länder* were able to reclaim the land which had once had seignorial status (the so-called *Domänen*, or demesnes). Only Sachsen-Anhalt availed itself of this right and so became the custodian of a substantial amount of land. The future of land expropriated from large landowners was more problematic. The solution to this problem, still contested by some, was not to honor claims on any land expropriated during the Soviet occupation, that is, between 1945 and 1949, while allowing claims by Jews whose land was expropriated by the Nazis and also those by individuals whose properties were expropriated by the GDR regime after 1949. As we shall see later, the state gave preference to descendants of the erstwhile owners in leasing land they had once owned. Such leases may lead to eventual ownership.

Rights to Compensation and to Decide on the Future of the Cooperatives

In addition to the restitution of full rights to land that members had brought into a cooperative, they were able to make other claims on the cooperative's assets. In June 1990 (revised in 1991), the Law on Agricultural Adaptation *Landwirtschaftsanpassungsgesetz* (LAG) was instituted, giving LPG members a number of rights they had not had under the old laws governing LPGs.[2] Thus, as we have already seen, while the old law recognized private ownership of land brought in by its members, inheritance rights in the GDR were vague. For example, the obligatory initial investment could only be passed on to heirs if the latter had also become members of the cooperative and continued to work there. The first change was the stipulation introduced in March 1990, that heirs had a right to an amount equal to the value of the initial investment in the LPG, payable in ten equal annual installments. The LAG went beyond this law, recognizing the right of the members to the capital of the LPG and, as a correlate, a voice in decisions regarding the LPG's assets. Since the united Germany does not recognize the LPG as a legal form, they were directed to ask the full membership to decide on the future of the LPG. That decision, which required a two-thirds majority,[3] had to include the value of each share in the farm and the number of shares to be allocated to each member. Each member had the right to at least one share. While individual members could not contest the decision reached in the meeting, they could question the amount allocated to them (be it because they felt discriminated against compared to other members or because they believed that the LPG's assets were undervalued) and request additional compensation in cash. Each member's share of the capital was calculated in accordance to the amount of assets other than land initially brought in by each member, the amount of land, and the number of years of service. The individual assets were to be figured according to their original value plus 3% interest (the rules did not stipulate compounding the interest). A rent amounting to 3 DM per *Bodenpunkt* per year was allocated for land placed at the disposal of the LPG. Finally, compensation for past service was to be calculated according to the ratio of the total amount the LPG paid for wages since its inception to the total amount for individual assets, rent, and interest. Thus, ownership of capital and land was given priority over labor. While members were expected to leave the amount of compensation due to them in the cooperative in the form of shares, if the membership decided to continue joint production, those who found work elsewhere or started their own farm were to be paid in full within a period of five years.

The members' options were limited to dissolution and the recreation of a variety of new individual and corporate farms, or to reconstitution as a western-style cooperative that rents the land from members and nonmembers. Some chose to liquidate and distribute their assets. The choice to liquidate collectives had the advantage that enterprises formed by the members after dissolution were exempted from carrying any remaining debt from GDR times. In contrast, LPGs that chose to reorganize had to carry such debts, which due to the artificially high evaluation of the GDR mark when currency union occurred, meant a de facto 300% increase of the debt (Laschewski 1998:55). On the other hand, liquidation was also fraught with danger. For even if its assets had a high appraisal value, the amounts actually distributed to the members were often negligible or nonexistent. One cooperative manager described the case of a collective in a nearby *Kreis*. "That collective, which farmed almost 7,000 hectares of land, was not in any worse shape than we are. But the manager, who was 55 years old, said, 'I can manage with the old age transition money (*Altersübergangsgeld*). Even though it is not much, my future is secure.' So he said he wouldn't do it. The others also refused. No one dared to take on the challenge. The liquidators appeared on the scene right away. They took over and within a few years more than 7.5 million marks were dissipated. Not a penny is left. The liquidation is still not over. The farmers had expected a lot from the liquidation. They had heard how much there was supposed to be and thought they would receive a corresponding share. They are amazed that nothing is left. You know how those things work. I know only one example like that. But there were many similar cases of enterprises that could have been continued in the interest of the cooperative members. In many cases some money was paid out. . . . But each member could have said, 'Even though I am retired or have taken early retirement, I will remain in the cooperative and will see to it that the assets grow and I receive a share from its operation.'" Ultimately, one-fourth of the collectives either chose dissolution from the start or, by 1995, had failed in their attempts to make the direct transition to cooperatives or other corporate forms and were forced to liquidate (Luft 1997:11).[4]

Most collectives chose to reorganize themselves as cooperatives according to West German law after an interim period as a partnership in which the future managers acted as limited partners. At the same time, various types of individual farms, limited partnerships, and other corporate forms emerged. But, in spite of generous subsidies available for all private agricultural ventures, special incentives promised to smaller farmers, and particular preferences conceded to local farmers in gaining access to state-owned land, relatively few local farmers took the step to become independent in the area studied. In eastern Germany as a whole, 21.7% of the land was cultivated by individuals in 1996 and an additional 15% by partnerships, often formed by fathers and sons. Only 7,029 new full-time farmers established family farms and an additional 17,985 became (or remained) part-time farmers (see Table II.1).

TABLE II.1
Juridical Forms of Agriculture in the New *Bundesländer* (1996)

Juridical Form	*Farms Number*	*Farms Size in has*	*Percentage of Total Surface*
Individual Farm	25,014	48	21.7
Principal Occupation	7,029	123	15.6
Supplementary Occupation	17,985	15	4.9
Partnership	2,465	338	15.0
Common Law Partnership (GbR)	2,291	363	15.0
Other	174	49	0.0
Juridical Entities (Private)	3,249	1,079	63.1
Cooperative	1,293	1,425	33.2
Limited Enterprise (GmbH)	1,432	826	21.3
GmbH & Co KG *Kommanditgesellschaft* (mixed partnership)	355	1,126	7.2
Stock Company	54	1,369	1.3
Other Juridical Entity	115	72	0.1
Juridical Entities (Public)	115	86	0.2
Total Number of Farms	30,843	180	100.0

Source: Reproduced from Thiele 1998:41.

In the Bernburg area, these figures were even lower. In particular, we encountered few part-time farmers, a fact confirmed by the statistics. In addition, 2,456 partnerships were instituted.

The remaining chapters of this book will be dedicated mainly to the latter two options: the reconstitution of LPGs as western-style cooperatives and the formation of individual farms.[5] However, we will return to the dissolution of state farms and LPGs when we discuss the formation of private farms, since these (but not the modern cooperatives) often obtained land freed by the breakup of such farms.

In this transitional period, then, a major reevaluation of the worth of goods produced and land was undertaken. At the same time, support structures were altered and a state system of distribution was replaced by a western-dominated system of supply and an, in part, artificially created system of demand.

Of all these changes, the process of reprivatization of land had the most profound and vexing consequences for former members of the collectives and their managers. It meant a rethinking of the value of things and the worth of relationships. The managers felt betrayed by their members' loss of communitarian values in their willingness to consider removing their land from the collective and even forcing its dissolution. They also felt disheartened by the ease with which consumers fell prey to western products. Finally, they felt betrayed by the west because of the frequently unscrupulous behavior of early representatives of western business and professions. They had to convince the membership and the state of the long-term benefits of joint farming and the dire consequences of dissolution, including the loss of jobs and previously added value.

It has often been stated that the social movements of the early transition period of 1989–90 were completely drowned by the subsequent acceleration of the reunification process, which amounted to a de facto annexation or colonization of the east by the west. We would argue that, in the case of agriculture, this view may not be entirely accurate. In spite of the voices that demanded the total dismantling of agricultural collectives, their reformulation by the transitional GDR de Mazière government was largely adopted by its pan-German successors. The efforts of the New Forum and other social movements thus did lay the basis for an at least somewhat more inclusive hegemony, a hegemony that allowed actors in agriculture to retain, in modified form, certain ideas and practices from the communist past that they cherished. Meanwhile, a few individuals who dared to strike out on their own searched in their immediate, but mostly in their distant precommunist past, for models to aid them in their endeavor. They were soon joined by westerners who also had much to learn in their new setting.

CHAPTER FIVE

The Cooperative Solution

Mulling over the difficulties of transforming an agricultural collective in Eastern Germany into a cooperative conforming to western German law after the reunification of the two Germanies in 1990, the manager exclaimed, "Let's not fool ourselves. The attitude today is still to get rid of the cooperatives if possible. They are not welcome under the present political circumstances. I don't know why this is the case. Probably, they prefer it if Mr. X has 10,000 hectares as a private individual and has leased the land away from a hundred or a thousand farmers, because they can't subsist. They probably think that that is a more peaceful solution than to say that there is a cooperative with fifty farmers that has also leased additional land and has provided the daily bread and wages for them. Somehow there is the idea that that is not right, that a cooperative is something sinister.

"I am of the opinion that Raiffeisen had the right idea already in the past century. You know as well as I do that no political system has succeeded in realizing the true will of the people. . . . I could have established a partnership (*Gemeinschaft bürgerlichen Rechts* [GbR]) or a trading company (*offene Handelsgesellschaft* [OHG]). I had many options. I could have done it privately as well. But [one of the reasons that I didn't was] that I was angry that people said that cooperatives had no right to exist. [I could not understand] this absolute judgment. If a GmbH has the right to exist, what difference is there? In a GmbH we would have been involved [with our capital], but so what. Just because the word 'cooperative' has negative connotations? If no one can prove to me [that a cooperative is wrong], then I will fight to my death that cooperatives have a right to exist. That is, if it is managed according to the true principles of cooperativism.

"When I really push [the lawyers] and demand an answer, their ultimate response is, 'Well, that's just the way a market economy is.' With that the question is settled. They remind me of Erich Honecker and his Soviet bedfellows in GDR times. When they no longer had any ready answers, they would ask, 'Are you for peace or not?' Of course we were all for peace and so we would have to do things differently. Well, if a peace policy was only possible because we subjected ourselves, I can still understand that, because peace is something magnificent. It's worth making a sacrifice, but if I have to submit myself to the market economy, just because it is the market economy, I can't see any reason for that. It isn't the market economy but the forces of a free market that drive me and I don't want that. That means giving up my autonomy and my identity, which I refuse to do. . . .

"I know that at present and in the foreseeable future no farmer can build an existence with 5 or 10 hectares of land. What I do know is that out of ten farmers with 10 hectares each one has a chance. But if he does it alone, he will also only survive for a short while. But when I tell five or six farmers in a community of thirty or forty farmers to form a cooperative, that creates a political force that provides bread and work and also gives support to local interests.

"In contrast, those who found a GmbH have no interest in the locality. I think that at least some of those who own the means of production should be able to be what they want to be, namely, farmers. They would also have a chance to do that in a partnership, but in that case each one of them would have to lease land from ten others before joining the enterprise. Why take such a circuitous route? For tax reasons? Or because of prestige? I doubt that. I know that this runs counter to the beliefs of many politicians, but I don't care. It is my opinion that cooperatives join people who are active in the same type of production to work together and have joint responsibilities. It gives those who share the same beliefs a chance to continue living. That doesn't occur in the GbR. They would sideline everybody, and I would be king. We give such great importance to property ownership. Well, aren't [cooperatives] precisely giving the owner a chance to do something with his property?"

The foregoing quote gives a sense of the ideological commitment managers of cooperatives still felt to the ideal of cooperativism. Although often in a less pronounced form, these managers invariably spoke of the need to maintain the interests of as large a collectivity as possible as the major rationale for transforming collectives into cooperatives rather than adopting some other enterprise form. The quote also portrays the embattlement that the managers felt in the early and mid-1990s and continue to feel today at least to some extent.

Legal Options

An LPG could reconstitute itself as a common law partnership (GbR), a partnership with limited liability (GmbH), as a shareholding company, or as a cooperative (*eingeschriebene Genossenschaft*). It could also reconstitute itself as several juridically independent companies with agricultural production as a cooperative and other activities as GmbHs or other corporate forms.[1] Out of the 3,844 LPGs that existed in 1989 in Eastern Germany, 3,126, or more than three-quarters continued in some corporate form in 1995. The preferred forms were cooperatives, of which there were 1,315 cultivating 34.2% of the agricultural land, and GmbHs, of which there were 1,417 farming 21.6% of the land (Luft 1997:10–11). In the area studied, the collectives that survived the transition all became cooperatives.

Eingeschriebene Genossenschaften, also known as *Raiffeisengenossenschaften*, date to the mid-nineteenth century and are named after Friederich Wilhelm Raiffeisen, who promoted the idea of cooperatives. The legal framework dates to the law on cooperatives of 1897. Patterned after earlier efforts in England to improve living conditions of the poor by enhancing their purchasing power through group purchases,[2] the original founders of Raiffeisen cooperatives never had the specific conditions prevailing in industrial production, much less agricultural production, in mind. They designed them to make it difficult for cooperative members who leave to decapitalize the cooperative by making it impossible for them to lay claim to more than the amount they initially brought into the cooperative, unless the entire cooperative is dissolved. However, if the cooperative's liabilities exceed its assets when a member leaves, he or she may receive little or nothing back. In addition, in some of the cooperatives, an individual can only remain a member while he or she is actively involved in the cooperative, or, in a common variant, during retirement in old age.

While the first rule surely makes sense in a purchasing cooperative, where individual capital accumulation is not a major goal, and the second is not much of an issue either in a type of cooperative where members' labor input may be insignificant, these rules may have a very different effect on a producing cooperative. Unlike a purchasing cooperative, where there are incentives to keep the membership constant or growing, a producer cooperative may expect to shrink through rationalization in the workplace. This is particularly likely in an agricultural cooperative, where an expansion of the productive base is not likely.[3] The structure of this type of cooperative may lead to the concentration of the cooperative's assets in fewer and fewer hands. It is not clear why entirely new forms of cooperatives that would have served the interests of the LPG members better were not devised. Most likely, the farm managers were reluctant to openly espouse forms of cooperation not already well established in Western Germany and thereby jeopardize their chances of securing land leases from the *Treuhand* and bank loans. In addition, the laws underlying Raiffeisen cooperatives were familiar to the Easterners. Purchasing cooperatives of the artisans had partially continued to follow their precepts. Finally, cooperatives have a relatively broad latitude in formulating their statutes (see Laschewski 1998:41) and, therefore, given the time pressure to come to a decision regarding their future, LPGs may not have felt the need to experiment with entirely new forms.

The Cooperative Managers' Opinion of West German-Style Cooperativism

To what extent the managers of the new cooperatives were aware of this characteristic of Raiffeisen cooperatives is unclear. The managers pointed, instead, to the democratic nature of the decision-making process in cooperatives in which the owners of the land who are also employed in the cooperative have a greater say in the operation of the farm than they would enjoy under other legal forms. They lamented the hostility they sensed on the part of Western authorities against a form of democratic economic cooperation with deep roots in German history, a form that was first taken over by the communists, but then was corrupted by modifications that effectively deprived cooperative members of individual ownership rights to the assets they had contributed to the cooperative. They also felt that the cooperatives were able to conserve far more jobs than individual farms. Indeed, although the cooperative managers were cognizant of the need to rationalize agricultural production, they viewed job preservation as one of their major functions. The rationale for maintaining a cooperative form of agriculture is forcefully expressed by the quote of a cooperative manager given at the beginning of this chapter. Similarly, in response to our question of what the most difficult aspects of the transition after the *Wende* had been, one cooperative manager answered, "The knowledge that our continued existence as a cooperative was not regarded as politically desirable. Today that attitude has become less prevalent. After four years, the cooperative is still in existence. So, people are asking themselves whether they really still want us to go away. What would take its place? Of the 30,000 persons employed [in agriculture] in Sachsen-Anhalt, an additional two-thirds would be without employment. Currently, no one can afford such an outcome. I certainly believe that it would be wrong. . . . Everybody recognizes that it would be a crime to impose the kind of small-scale farming as it is still to a large extent practiced in Western Germany, but they don't recognize the system of private farming in the form of cooperatives. There is a simple reason for this. We are the successors of the LPG (*Nachfolgebetriebe der LPG*). That is true. But these cooperatives could just as easily be continued with one owner. That would be regarded as more acceptable. There are also GmbHs. The structure would be the same but with fewer partners. Let's say two or three, or four or five. That would not have worked well here. Three or four individuals would have come forth and would have said, 'We have the right and you can work. But we are the partners and can draw the overall benefit.' That kind of arrangement also exists in this area. But I am thinking of the way in which those [old] farms are being taken over by individual owners. Nearby there is a 4,000- hectare LPG that was taken over by [a well-known West German]. He came in with the money and took it over. That didn't happen with us. But that is regarded as more acceptable than a cooperative like ours. But I trust that it won't take long before people will say, 'The old times have gone. Today those things (i.e., the communist structure of cooperatives) are no longer in existence.' Everybody can see how these enterprises are now constituted. The registry court does not recognize any LPGs, only enterprises that were transformed according to the new law. [The perception] will change one of these days."

The Transition to West German-Style Cooperatives: Process and Timing

While many LPGs waited until the last moment to dissolve the collective or change their status to an enterprise form recognized in Western Germany, some took steps to reconstitute themselves early on. One manager of a cooperative considered himself to be well prepared for the change. He had read extensively before the *Wende*, had listened to West German television and, although he was not allowed to travel to the West himself, he had kept abreast of developments in prices and the like in the

West by instructing those of his employees who could travel to inquire about such matters. He was also familiar with the manner in which West German cooperatives worked, for, as we have already indicated, some of their principles still applied to Eastern German cooperatives. "I said to myself, because of the tradition, we should establish a Raiffeisen cooperative. I always had the latest proposals for new laws. I always knew in advance what was coming. So, I already oriented my collective in 1989 so that we would found a Raiffeisen cooperative. Then we made the transformation and were registered as a cooperative already in 1991, before the law on agricultural adjustment was instituted. When they spoke about free elections, it was immediately clear to me that the GDR would not exist for long. So we quietly began to orient ourselves toward the joint German option, and adjusted our planning accordingly. I think that we decided in May 1990—unification only took place in October 1990—to become a Raiffeisen cooperative. We made the decision at a general meeting of the LPG without any votes against it, because no one knew what it was all about yet. We set a deadline of October 30 to join the new cooperative and to acquire shares. The new members acquired an average of 10,000 deutsche marks' worth of shares. [The amount was up to each member]."

Similarly another LPG manager began holding meetings concerning the future of the LPG as early as the end of 1989 or the beginning of 1990. "In the first quarter [of 1990] we must have had at least five to ten meetings. I must say that few LPGs were as active as we were. Many thought that we were experiencing a temporary change. They couldn't believe that things could be transformed radically, just like that. At any rate, those of us who were involved in management here felt that those who wished to establish independent farms should be able to quit the LPG and take their land and whatever else they might want out of it."

The New Leadership

The continuity in leadership exemplified by the foregoing example was the rule in the transformation of LPGs to cooperatives. All the cooperative managers we interviewed had held managerial positions in the collectives out of which the cooperatives were formed. According to a study undertaken in the same year as our own, 64% of a sample of eighty-one cooperatives in Sachsen-Anhalt continued to be managed by the same manager (Laschewski 1998:131, 203; see also Koester & Brooks 1997). However, many cooperatives are not headed by the former LPG managers. Many were close to retirement age or took the option of retiring early, while others did not dare to take the step to capitalist agriculture. While the number of candidates for top leadership positions was often low, the political ferment of the months preceding and following the fall of the Berlin Wall also had its repercussions in the LPGs. Not only were managers elected, but the electoral process became more meaningful than in GDR times, when preselected single candidates were invariably voted in unanimously. For example, in one instance, the LPG manager retired because she had reached the age of 65 in 1990. The designated heir to the position was a younger man who, like his predecessor, was a member of the Communist Party. In addition, he had joined the LPG quite recently and was not well known by its members. The present head, a former professor who had lost his position at a university for political reasons and had held a lower managerial position in the LPG, set himself up as an opposing candidate. As he explains, "Before, something like that would have been impossible." Nevertheless, the response to his announcement was more in keeping with GDR political culture. "Thereupon, [the designated candidate] retracted his candidacy," the consultant continued. "Now, there was suddenly only one candidate again, the way it had always been, and so I became the manager." Another cooperative manager, who had formerly managed the livestock LPG that became a part of the new mixed farming cooperative, prides himself for not only having held elections by secret ballot already before the *Wende*, but having instituted elections of the executive council, the board supervisors, and the manager, while the more general practice, suggested by the cooperative law, is for the board of supervisors to appoint the executive council that in turn elects the general manager every three years. Laschewski (1998:132–134) found that in the three cooperatives he investigated intensively, leadership qualities and loyalty to a particular interest group, in this case members who owned land in the cooperative, were regarded as more important in selecting a manager than occupational competence. Further, older respected individuals had a major influence on the person chosen, and they often chose persons who had already been selected for further training in GDR times (Laschewski 1998:133–134).

By 1999, the managers of three of the cooperatives we revisited had retired or were about to retire. The succession presented few surprises. One of the new managers had acted as submanager for many years. His father, a refugee who had joined the cooperative in 1953 when it was not mandatory, had already managed the cooperative, while the son and present manager was working in the cooperative and studying agriculture. After his university studies and before becoming general manager, he had been in charge of feed production and then of the entire production. In another case, the manager of the livestock unit took over the reins of the whole cooperative. And in the third case, the present manager was training a 30-year-old woman for the job. There appeared to be no question in his mind that his choice would be honored by the board when he retired in a few years. Interestingly, two of the new managers are women. We know of only one collective in the area that was managed by a woman during the GDR, and she took early retirement during the transition period. None of the managers we interviewed or had heard about in 1993–94 were women. It remains to be seen whether these cases of female management represent a greater openness toward women in managerial positions in agriculture.

A question that we did not broach during our first fieldwork because it seemed indelicate to do so but did come up during three interviews in 1999 was why our manager informants had not decided to become independent themselves. One of these three managers had already told us in 1994 that a GmbH was, perhaps, a more appropriate enterprise form than a cooperative because managers had more authority. In 1999, he repeated this opinion and went on to say that if he could start over again he would either become independent or found a GmbH. The second informant, a man who had recently succeeded a retiring manager, said that he was poised to become independent and had even purchased a tractor, but the bank would not back him because he only had 5 hectares of land while his house was not substantial enough to be used as collateral. For him, and presumably for other cooperative managers as well, running their own farm would have been considerably more lucrative. He was the only informant to tell us what he earned, only slightly over 5,000 deutsche marks per month. "Our accountant," he said, "feels that is too little. We can only improve our standing because everybody receives 5% more each year and 5% of 5,000 is more than 5% of 2,000. But that is the nature of a cooperative. That's the problem. Everybody is an owner." Finally, the third informant did become independent when he had the chance to take over the demesne that was formerly rented to the cooperative that he had managed. Although he was unwilling to give any figures, he did concede that he was earning more than when he managed the cooperative, not to mention the fact that a beautiful old house came with the demesne. We expect experience, loyalty, and recommendations to continue to be important, but gender and age might change in time.

Carrying Out Compensation

The manner in which the mandates of the Law on Agricultural Adaptation have been carried out varies from cooperative to cooperative.

Some cooperatives negotiated an extension of the deadline for paying the compensation. These negotiations were time-consuming and nerve-racking. As one cooperative manager explained, "When one changes the period to pay compensation from five to ten years, no one is there to advise you, you have to be your own advisor. In most cases, I talked to each individual in person. I went to their homes and explained the way things hang together and there were very few cases where there was no resolution. They had a lot of understanding, but it was also a war of nerves. I had to deal not with five persons but with 400 individuals and their families. When there was no understanding and there were negative feelings toward me, I had to tell the individuals that we were in the same boat and that if we had to go through bankruptcy proceedings they would not obtain their money either. Most of them understood that." According to another cooperative manager, "We have to pay and are paying 2 to 3 million deutsche marks within five years. Since we were starting afresh, we told our people that although payment was assured, we would appreciate it if we could have four additional years, that is, until 1999 [to pay them]. We talked to each one individually. We told them that they could decide whatever they wanted. If they decided against membership, they would be paid within five years. Many have sons and daughters who work for us or work for us themselves and those, of course, said that the enterprise had to remain viable or they wouldn't receive anything at all. So they signed the agreement. We failed to convince [only] four nonmembers. They will be paid in five years. Last spring, since things went well financially in the past three years, we told those who had signed the agreement until 1999, that we would be willing to pay them 35% of their share in cash, if they would forgo the balance. Almost 40% of the members took advantage of the offer, mostly those who were around 70 years or older. They could still see how their grandchildren were happy about receiving the money. Those who were still in good health could make good use of the money to make life more pleasant. A large number took the money and put it in a savings account. Last year, interest rates, which hovered around 6.5% to 7.5%, were attractive. They said, 'I will leave the capital and spend the interest.' . . .

"It worked like this. On July 1, 1990, we had assets worth some 5.5 million. Of that, we had to subtract our old debts, that we will have to pay sooner or later, leaving 2.8 million. Initial inventory contributions from our members added up to 450,000 to 470,000. That left 2.3 million. Of that amount approximately 80% was allocated to the use of the land, 10% for the years worked and the cooperative was allowed to keep the remaining 10%.[4] The 80% allocated for land use resulted in

2,438 marks per hectare. So a person who brought in 10 hectares of land will receive 24,380. For the years worked—we will make the payment this November—the amount will be 70 marks per year worked, so that if someone has worked for us for 30 years, he or she will receive an additional 2,100 marks in compensation." While, in this case, none of the payments appear to have been made at the time the new cooperative was formed, in other instances a considerable part of the payments were made in the first year. Thus, in one cooperative, all members over 65 years of age received their share immediately. In another, two-thirds of the total sum owed to the members had already been disbursed by the end of 1993.

Some cooperative managers concede that it was, in practice, impossible to make equitable determinations about the amount of compensation specific members should receive. As one manager described it, "It was almost impossible to determine who had brought in equipment and who had not. A large proportion had not. We saw that from the individual protocols, and from the many Type I cooperatives (i.e., those who had not pooled their animals). Those were the small islands [in a sea of Type III cooperatives] that later joined the larger enterprises which were then included in the large LPG. Some of these assets were brought in jointly, while some Type I farmers had to make an additional payment [to make up for the difference between the appreciated value of the original contributions of the already established members of the larger LPG and the appraised value of the assets of the Type I LPG]. Those [original] individual payments were fully recognized by us. In terms of the land, we made a uniform determination and, in addition, members received payment for the fact that many had already planted their fields when they originally joined, even though some—or a majority, who knows how many—had already received compensation."[5]

Major differences in compensation resulted from the divergent histories of the LPGs that were joined to form the new cooperatives. As we have seen, in most instances, crop-growing LPGs were much better off in GDR times than livestock-raising LPGs. And yet, due to the system of syphoning off excess profits, agreements by crop cooperatives to forgive debts incurred by livestock cooperatives, low-interest loans, and wage controls, the members of a profitable LPG were not very substantially better off than those in one that was heavily in debt. These differences now suddenly became important in determining compensation. They could, for example, result in a situation where two individuals, who once belonged to the same LPG before it was regrouped with another LPG and divided into crop- and livestock-raising LPGs, would receive vastly different amounts in compensation for land used by the same crop cooperative. In one instance, a livestock LPG with debts amounting to 2.5 million deutsche marks was joined with a crop LPG where the members expected payments totaling over 2.5 million in addition to the 1 million already paid out, the former members of the plant LPG received 160 deutsche marks per hectare, for every year the land had been used by the LPG; those of the livestock LPG with land in the same crop LPG remained empty-handed.

The new cooperatives also varied with respect to the minimum entry capital they expected from their members and whether that amount had to be paid up front or merely deducted from expected compensation. In one cooperative the minimum share was 2,500 deutsche marks, while in another no minimum level appeared to have been fixed, but members purchased shares for an average 10,000 deutsche marks per person.

Cooperatives still feel beleaguered, in spite of having made what they feel were very considerable efforts at keeping everyone's interest in mind in the transformation. One manager remarked philosophically that although all the members had approved the compensation proposal, which had also been examined by a lawyer and by the government of the *Land*, "If any member has the impression that something was not quite right, the authorities would judge that he was in the right rather than the new enterprise, 'that successor of the former LPG.' But we have to live with that. Until now we have held our ground and have succeeded in accomplishing all this without quarrels or lawsuits. We also gave recognition to the fact that 50% of the land was already under cultivation when our present cooperative took it over. The owners will receive compensation for this when their present lease contracts have run out." As we shall see in chapter 7, the *Wiedereinrichter* have a less benign view of the compensation process.

By 1999, compensation of the members of the former collectives we studied had largely been achieved. One of the cooperatives studied had already paid and others were in their last years of paying off their debts to these members and were, therefore, looking forward to having more disposable resources to invest in their operations.

Impediments: Access to Land

Farmers who opted to reorganize as a Western-style cooperative, which in most instances was clearly preferable to dissolution, faced numerous pressures that led to a decrease in farm size. First, cooperatives were urged to reconstitute themselves into smaller units roughly corresponding to those formed in 1960, during the first wave of consolidation and forced membership. This essentially entailed allocating each livestock cooperative a corresponding share of the land cultivated by the crop cooperatives. Most cooperatives accepted the advice of the Western experts that the former LPGs were too large for rational operation and proceeded with the division into smaller units which, as we shall see later, reintegrated crop and animal production. Some cooperative managers later regretted the decision to create smaller units, for, as it would turn out, this choice made them even more vulnerable to the further erosion of their already rapidly decreasing land base. In the words of the manager of one cooperative, "[The division of the LPG into smaller units] was dictated from above. The minister of agriculture said, 'They are all too large and have to be divided.' We decided that it would be to our advantage not to swim against the political current. However, it soon became apparent that we are always swimming against the current anyway." Not all cooperatives chose to reconstitute themselves into smaller units. One of the most successful LPGs in the area, which is still considered a model cooperative today, chose to maintain its former size and, as we shall see later, even many aspects of its former structure. In two other instances the livestock cooperatives could not come to an agreement with the crop cooperative with which they had formed a *Kooperation*.

It should be noted that not all managers of cooperatives regarded the splitting of crop collectives into smaller units and their unification with livestock units, that were often organized on a territorial, that is, village basis, as a solution that was imposed from above. Some regarded the *Kooperationen* as too large and unwieldy to form the basis for the new cooperatives. Particularly in regions with a historical preponderance of small-scale farms, members of collectives pushed for the reinstitution of the situation before the village-based collectives were united into larger units and crop and livestock production separated (Laschewski 1998:108, 124).

At the same time that they were shrinking in size as a result of the restructuring process, the new cooperatives were discriminated against by the authorities with respect to the allocation of land. While the authorities could not dictate what the members could do with their land, they could restrict access to state land and demesnes belonging to the *Land*. The *Treuhand* gave preference to local farmers and Western individual farmers over cooperatives, while the latter were, at least for a time, excluded altogether from renting land belonging to the *Land* of Sachsen-Anhalt on a long-term basis. As one manager of a former collective and present manager of a cooperative explained, "We made a request [to rent land belonging to the *Land*] citing our planned investment of 1 million marks. We will not get the land. [A cooperative] is regarded as a form of socialism. I take a dispassionate view of the situation. We did not win the war. They (the state) have to find a way to ensure that the Western farmer does not lose out in the process of German unification; that is why they continuously create impediments for the agricultural cooperatives. It is possible to operate economically under the present circumstances. The difficulties are of a political nature. . . . To be sure, things were just as bad or worse when the collectives were originally formed. Those who refused to join also had problems. I come from a farming background and know that the pressure on individuals to join was very strong. It included economic pressure such as [differentials in] veterinary fees, fertilizer costs, and things like that. I see definite parallels in the present situation, except that the pressure now goes in a different direction. We have to adjust to this situation."

Other cooperatives are still hopeful that the *Land* will accept one or another of the proposals they have submitted. One cooperative manager declared, "We submitted two proposals already seven or eight months ago, but the gentlemen and ladies [in charge] are unable to come to a decision. First, we are interested in purchasing the buildings. That would be the most logical solution. We are sitting in them and have maintained them over the years. The buildings are in good shape and so we are of course interested in harvesting the fruit of our efforts ourselves. In addition, we would sign a twelve-year lease for the land. We also submitted a second proposal if only a private individual rather than a cooperative is eligible to make such a purchase. We would then suggest a private person who lives here and has worked for us in the past and has a degree in agriculture, but no longer works here at present to make the purchase. We would then act as guarantors and form a GbR or a GmbH with that person. But our dear friends at the *Land* government can't make up their minds, and so everything remains uncertain. As long as there is no decision, we can only rent the land from year to year." Still another cooperative manager was much more sanguine about his hopes to be able to take over a crumbling demesne himself and cultivate the land that pertains to it in conjunction with the cooperative.

All cooperative managers are still trying hard to hold on to the land. They argue that they would have to abandon major parts of their livestock operations, which would add substantially to unemployment, since private farms are much less likely to engage in this capital- and labor-intensive activity. One manager commented sadly, "The trend has been to take away as much land as possible from the cooperatives. I don't know what they want to achieve by doing this. It leads to the loss of workplaces and to worse social conditions. We have already reduced livestock production to a minimum. We want to start raising bullocks again because we believe that there is some profit to be made there. But our plans are uncertain. If there is a year in which few sugar beets are harvested and the sugar factory decides to raise the price of pressed sugar beet [fodder] to 4 or 5 marks, then I will have to think long and hard whether I still want to continue raising the bullocks and somehow weather [the price increase]. Under these circumstances it would help me a lot if I had a bit more land, which would give me the extra cushion to compensate for the loss. But if I have less and less land at my disposal, then I will get to the point where I will have to get rid of any activity that is the least bit risky. The government of Sachsen-Anhalt and its agricultural minister deeply regret the fact that livestock numbers have declined so precipitously, but they themselves have been the main cause for this decline, by reallocating the land held by the *Treuhand*. . . . I have tried to make this argument wherever I could, but I believe that politics is only rarely practiced according to rational criteria. Most of it is done emotionally, from the belly. Feelings dominate." The loss of the demesne land would indeed be a significant blow to many cooperatives in Sachsen-Anhalt and particularly in the study region, where demesnes were once prevalent. In one instance, it would entail a further reduction of 300 hectares of land, or 20% of the surface cultivated in 1994; in a second case 350 hectares, or 19.4% ; and in a third case, where there are two demesnes with a total surface of 630 hectares, 37%.

The *Treuhand*, or the *Boden-Verwertungs-und-Verwaltungsgesellschaft* [BVVG] (Land Administration Agency) as its successor is called, has been somewhat less biased against the cooperatives than the *Land*. However, as indicated earlier, it gives preference to former estate owners whose land was expropriated during the Soviet occupation in leasing the land it holds; Eastern Germans who are trying to reestablish individual farms; and to a lesser extent, Western farmers who are not former estate owners. The uncertainties surrounding land ownership; vacillation about who should be given preference; the plans of the *Kreis*, which also has a voice in such decisions; and often just sheer incompetence was creating long delays in signing long-term leases, without which the farms have greater difficulties planning for the future and in obtaining bank loans. As one cooperative manager commented, "We have leased almost one-fourth of our surface from the *Treuhand*, mostly for only one year and now for two years. It is quite possible, that in the next few years some 10% to 15% of the land will still be taken away. If we can keep the land, then we can take full advantage of the large machines [we have purchased]; if the surface is significantly reduced, then we have made a bad investment."

In the time period between our two field trips, the amount of land leased from the BVVG had indeed further decreased, partly because it was leased to individual farmers but mostly because of the construction of a limited access highway. Thus, one cooperative has lost 310 of the 1,560 hectares of land it had held in 1994. Concerns about the environmental impact of such roads has led to the institution of a program whereby the state acquires three times the surface needed for the actual highway. The balance is set aside for reforestation programs and in order to enlarge natural parks. While the state also purchases land from individuals, the path of the highway often leads through land over which the state already exercises control. In this manner, the cooperatives we interviewed lost between 100 and 500 hectares of land each. One of the cooperatives we revisited also lost access to demesne land, which was allocated to a private individual—the former manager of the cooperative—and to a GmbH which was also founded by members of the cooperative. On the other hand, at least one cooperative we revisited was continuing to rent demesne land on a two-year contract basis and the manager was not too worried that this arrangement would end soon. Meanwhile, the BVVG continues to favor individual farmers when it reallocates the land in its possession. The BVVG even appears to have broken a longer-term lease it had signed with one cooperative in order to allow a Western German to purchase it and begin using it immediately.

At the time of our second visit, in 1999, the BVVG had already instituted a plan to sell off all its land by the year 2001. The idea was to sell part of the land at a rate substantially below its current market value to farmers who had initiated operations before the end of 1990, with preference given to individual farmers. Unfortunately, for our informants, the program was suspended due to the intervention at the EU level of individuals who did not stand to benefit from the program. Although, as we shall see, some of the individual farmers were already able to take advantage of the program before it was suspended, none of the cooperatives revisited were able to do so. In 1999 all land sales by the BVVG were held in abeyance.

Leasing from Individuals

The principal source of land for all cooperatives are individual members of the LPGs out of which the cooperatives were formed. Members of the cooperative who own land within its boundaries are expected to sign leases with their cooperative alone. Churches are a second source of land. However, the latter have often been reluctant to lease their lands to cooperatives, for they regard them as successors to their longtime enemies. In 1994, some had already signed long-term contracts with individual farmers and some made short-term contracts with the cooperatives. The renewal of these contracts was still uncertain. A third source are the thousands of individuals who fled to the West, as well as former LPG members. The process of finding the owner of each of the hundreds of plots of land out of which the LPGs were formed has been laborious and time-consuming. For example, in one instance, a cooperative had to find and sign leases with 169 owners of 469 parcels of land amounting to a total of 700 hectares. Of these almost 100 owned less than a hectare of land. Some of the cooperatives computerized their leasing operations from the start—in fact, sometimes even before reunification—while others were only beginning to make the switch to computers in 1994. As one cooperative manager explained, "The first year was terrible. People would appear and say that they had a piece of land here and we would have to go to the registry to see whether the claim was correct or request proof from the person. Of course, it was easier for us if they brought the proof. Then we would ask them if they wanted to sign a lease agreement. We also wrote to many individuals whose addresses we were able to get through relatives and acquaintances." Other sources of information were the land registry, the office for pending property issues, and older community members. Some of the owners lived as far away as Australia and South Africa. "Many don't even know that they own a hectare or two of land here," the same consultant continued. "They are often grandchildren of the original owners. They tend to have unrealistic expectations about the value of their property. For these reasons the lessor rather than the owner of the land must pay the real estate taxes. The state would only be able to collect half from the owners, if anything at all, for it would be incapable of finding all the owners even in ten years." For parcels of land where the cooperatives were unable to find the owner, they have signed leases with the *Kreis*. The court then appointed a custodian to manage the account. If the last heirs had died more than thirty years earlier, the cooperative could submit a request for a transfer of the title. In the example cited above, no owners were found for eighteen out of the 469 hectares of land.

Uncertain about the value of their land, about potential industrial development in the area where they own property, and, perhaps, their own plans to become independent farmers, the owners have often preferred to sign only short-term leases, some for as little as two years. To encourage the signing of longer-term leases, the cooperatives established rates according to the number of lease years. Thus one cooperative paid

2 deutsche marks per land value point for leases that can be terminated with one year's notice, 3.20 marks for three-year leases, and 4 marks for twelve-year leases. At least in some instances, there were provisions to adjust leases when the cooperative's economic situation had stabilized. A further mechanism to adjust rates is to tie them to the price of wheat. In 1994, a typical average rate per hectare of land was 269 to 320 marks. In addition, the cooperatives paid some 56 marks per hectare for land taxes.

The competition for land with individual farmers was fierce and sometimes dirty. A poignant example is the following interview with a cooperative manager. "We had 4,200 hectares before the *Wende*; afterwards we were left with 1,000 hectares less, although none of our members became independent. But we do have three Wessies in our territory." Pointing to a parcel of land from the car in which we were inspecting the cooperative, he explained, "This place is where in the 1950s a group of farmers with farms of less than 50 hectares had abandoned their land. These farmers have now regained this land, some 200 hectares altogether. They have not leased the land to us but to Westerners who continue to live in the West. One of them abuses the land more than the other. . . . After the *Wende* all sorts of people came from the West to buy land and old houses. Our people trusted them and gave them power of attorney. They then disappeared with the money. . . . In B there are three well known cases of such theft. In one, the former owner has had to pay 16,000 deutsche marks in lawyers' fees, but he hasn't received any of his money back."

The unconscionable scramble for land is also portrayed in the following statement by a cooperative manager. "In the vacuum created at the time . . . no consensus could be found . . . so into the vacuum came all sorts of folks from the West who caused fear among the people. These folks really didn't have the right to lease land but they did so anyway. Legally, the old leases remained valid. No one had abrogated them. But who wanted to get into a hassle with those people?"

In order to prevent competition among themselves, cooperatives operating in the same area came to some agreement regarding lease conditions for both privately owned land and state land. With the *Treuhand* finally giving indications that it was ready to sign twelve-year leases in 1994, such practices as using up residual fertility, which could make it difficult to restore the fertility of the land, will probably disappear.

Since some individuals had plots of land in two cooperatives or decided to lease their land to an individual farmer, cooperatives and individuals entered into exchange agreements, for which leasing contracts give specific permission. While such arrangements did not appear to present any difficulties when they were made between neighboring cooperatives, and the law mandated that cooperatives accommodate members who wish to establish their own farms, at least one cooperative manager conceded that he sometimes refused to enter into exchange agreements with Westerners to prevent the further expansion of their farms. We shall return to this issue in chapter 8.

As we have seen earlier, between 1994 and 1999, cooperatives continued to lose land to individual farmers. A cooperative manager explained the process as follows: "It happened like this. In 1990 we still had 180 workers and whenever it was socially responsible, some would leave, into early retirement or in other ways. Those who left and had leased their land to us might say, 'Now that my livelihood is secure, I can lease my land to my nephew who has become independent.' The uncle could also supplement his retirement benefits by helping the nephew during peak periods in the agricultural cycle. In this way we lost 40 to 60 hectares. Nevertheless it is unlikely that more large pieces will be lost."

In order to avoid such losses, both cooperative managers and individual farmers try to bind the lessors to the farm by wooing them with gifts at Christmastime and New Year's cards. As the same informant put it, "There are over 500 lessors who live all over the world: France, England, the most impossible corners. Few live around here, so it is not possible to invite them all to take a look at what we are doing. It's not like in the old *Bundesländer*. So we have to write them from time to time or at Christmastime. One doesn't really have close contacts with them. So there is no guarantee that one can continue to lease land from them. While, in general, leases are quite stable, there are fluctuations in the lease agreements. . . . There are also heirs who say, 'These people have taken our land from us and shouldn't have it any longer.' They maintain this attitude even if the present managers had nothing to do with those matters. We don't deny it. We say that we are the successors to the LPG. We want to be clear about it. But some persons have problems with it. On the other hand, we do offer advantages, also in terms of the way we cultivate the land. We continue to raise livestock and are thus able to improve the soil structure. Most of the individual farmers only raise crops, which eventually diminishes soil fertility." In addition to wooing landowners, all farmers are renewing leases, preferably for longer terms, many years before they actually run out and begin to pay higher rents immediately. Since 1990, typical annual land rents have gone up from 3 deutsche marks per hectare of 80 point land for a ten-year contract to 4.50 deutsche marks or 50%.

But the most secure way in which all types of farmers are trying to achieve more permanent access to land is by outright purchases from small-scale landholders. In 1994, few cooperatives had been able to afford agricultural land. The cooperative in our sample which probably has the largest proportion of land owned jointly, owned 40 out of the 1,300 hectares it cultivates. It was able to buy the land from a widow with no successors willing to start a farm of their own. It planned to purchase an additional 10 hectares in 1994, so that together with the land owned by its eight members it would have land assets amounting to 140 hectares. In contrast, the cooperatives revisited purchased between 50 and 180 hectares of land between 1994 and 1999. For example, the cooperative mentioned above now owns 150 hectares of land. Such purchases have accelerated in recent years as cooperatives are winding down their compensation payments to former members and, conversely, former members retire and wish to improve their standard of living by selling land and making a major purchase. Also, some owners may be selling because their unemployment benefits have run out and they need the money in order merely to subsist. Land is also frequently sold when it passes to heirs who find it easier to distribute cash than divide already small landholdings. Indeed, so much land is being offered, that some cooperatives find it hard to buy all of it. So they buy the land for which they have only short-term leases first in order not to lose it to someone else.

Some of our informants were surprised that individuals parted with their land so readily and even tried to dissuade them from such a move. As one informant described the land purchases, "In GDR times, land did not have the same meaning as it does today because it was cheaper.[6] At the time a good piece of land cost 3,800 to 4,000 marks per hectare. At the same time, people identified with their land. It was something that one did not just give away like that. People had acquired it painfully with their parents by working hard and saving. Now the young people wish to join the consumer society. For them land does not have an intrinsic value. Value simply means, 'I wish to have something for myself.' Mostly it is heirs who sell. The elderly don't sell. I explain to them that the rent from the land would give them the equivalent of the value of the land in twenty years. [But it is to no avail]." The fact that banks have tended to be wary of making loans to cooperatives, an issue that we shall explore in more detail later, has made it difficult for cooperatives to obtain mortgages for land even though the banks would not incur much of a risk. As one cooperative manager lamented, "We have purchased land from all the individuals who have offered it to us. But these large surfaces, 40 hectares (that suddenly came on the market) are impossible for us to buy. The banks say, 'My God.' In actual fact, the security would be there. It would be no problem to say, "Buy the land for 1.10 deutsche marks per square meter, it's a good price.' The interest on the loan would not amount to much more than we pay for the rent. The banks woke up when the land was gone, as always, too late."

An issue that will further affect access to land—in this case both in a literal and in a figurative sense—is the German governments' decision to institute a program of reorganizing land tenure (*Flurneuordnung*) that will entail making access roads to individual landholdings (after consolidating each individual's holdings into as few parcels as possible). While the owners applaud the decision because easy access will, in many cases, increase the value of a parcel of land, the program will certainly be a burden for most farmers, since the large tracts of land created in the GDR will again be interrupted by roadways. One cooperative manager declared disgustedly, "Each individual will have access to his or her parcel of land even though 90% of them have never cultivated the land themselves. But the law dictates it. They will make the trails anyway, even if it is sometimes idiotic. People want it because the land will be easier to sell, but they don't stop to think about the fact that it is going to cost money to keep the trails up. As a result, prices will rise further."

The Issue of Buildings

An issue that is, in part, related to securing of agricultural land is the acquisition of the land on which the cooperatives' buildings stand. During the GDR, LPGs built farm buildings wherever they wanted, without regard to landownership. Now, they have to come to terms with the owners of that land. Long delays could be expected if the land belonged to the state. And yet owning the land on which buildings stood was crucial for obtaining mortgages, and many loan applications were shelved until ownership of land and buildings was united. One frustrated cooperative manager complained, "I submitted a purchase proposal for some of the buildings three years ago. They are not processing it. They either can't manage or they are too lazy. We have had to resubmit the proposal three times, because they couldn't find it. As you can see, the conditions are not favorable." Even as late as 1999, some of the cooperatives had still not succeeded in acquiring all the land on which their farm buildings were located.

Purchasing buildings that once were a part of demesnes was possible only to the extent that the *Land* did not intend to rehabilitate them as part of a reconstituted old-style demesne. We will examine such attempts at resurrecting history in greater detail in chapter 8.

Set-Aside Programs

The problems of access to land are compounded by government-promoted set-aside programs and land use extensification. Within specified lower and upper limits, an agricultural enterprise can receive subsidies for fallowing, reforesting, transforming land into pastures to be used extensively, and using land for nonagricultural purposes. Additionally, in order to promote environmentally favorable agricultural practices and reduce production, the government subsidizes extensification of land used for products for which there is an overproduction, including beef and mutton, grain, potatoes, cauliflower, and tomatoes. In spite of the fact that the program favors farms with poorer soils—a disadvantage to farmers in the study area, where the quality of the soil is generally very high—the cooperatives in our sample have generally made use of at least the program to set aside land. In order to avoid farmers setting aside only their least productive land, the set-aside program stipulates that a given plot of land can only be left fallow for one out of every six years. Otherwise they must set aside at least 20% of their cultivable land.[7] The constraints of the land following program instituted by the European Union notwithstanding, this program has been a particular boon for the cooperatives. Since they were able to rent land at reasonable rates when they were formed shortly after the *Wende*, the cooperatives have been able to benefit disproportionately from this program, partially offsetting the disadvantaged position they occupy because of the rulings on the old debts (Koester & Brooks 1997:12–13).

The Carryover of Debt Incurred During the GDR

One of the most controversial rulings the German government made with respect to cooperatives was their treatment of debts incurred by cooperatives during the GDR. As we have seen earlier, while such debts disappeared when state enterprises were liquidated or when individual farmers took over liquidated collectives, collectives that were transformed into cooperatives remained burdened with them. The old debts, the *Altschulden*, constitute one of the most serious threats to the new cooperatives in Eastern Germany. On the basis of a sample of eighty-two cooperatives in Sachsen-Anhalt and Brandenburg, Wissing and Anz (1993:16) figure an average indebtedness carried over from the communist era of 3.88 million deutsche marks, but ranging from as little as a few 100,000 to 15 million deutsche marks. The old debts owed by cooperatives we studied ranged from no debts at all to around 3 million deutsche marks. Perhaps because the soils in most of the *Kreis* of Bernburg are exceptionally fertile, the average was much lower than that in the region as a whole. An analysis of a subsample of fourteen cooperatives in Sachsen-Anhalt showed an increase of 47% per hectare of land in the load of old debts resulting from land lost to other types of farms between July 1, 1990, the date of monetary union, and September 1992 (Wissing & Anz 1993:16).

The question of the old debts has led to acrimonious debates. The cooperatives contend that the LPGs out of which they were formed had often incurred these debts for projects that benefited the entire community, particularly roads which they were forced to construct and maintain. However, the debt incurred by the LPGs for these projects was transferred to the cooperatives that succeeded them, while all private farmers and limited partnerships were absolved from carrying any share. At the same time, the largely involuntary donations the LPGs were forced to make for other public works, such as the construction of dance halls and hospitals, could not be deducted from the debt. Detractors contend that such donations were merely a form of local and regional taxation. As proof, they point out that these payments could be subtracted from an LPG's overall tax liabilities. The cooperatives counter that these payments actually constituted a means by which the state decapitalized LPGs. They allowed the state to siphon off surpluses, leaving little for reinvestment. As one cooperative manager explained, "There was no cash. The enterprise had to hand over everything to the state. So if we built anything we were forced to take up a loan. We then paid the interest and the loans off to the bank. That could be deducted from our profit. The rest went back to the state. In principle, we simply treaded water, not more and not less."

To make matters worse, many of the investments that were made in the farms themselves are now useless. As we have seen, many of these investments were mandated by the state with little regard for the economic advantages to the LPGs involved. The mammoth irrigation project undertaken by one of the LPGs at the whim of government officials, who saw in it a way of making use of what remained of a failed project to filter water from the Saale River for industrial use, now lies in ruins because of the high cost of upkeep and prohibitive fuel costs to run it. With decreasing cattle herds and the almost total abandonment of sheep raising, even farm buildings that had been maintained up to the *Wende* now lie in ruins, their deterioration accelerated by vandalism by youths who have few recreational outlets and thieves who steal copper pipes and whatever salvageable items they can find. Similarly, many of the waste catchment basins, silos, fuel tanks, and other projects are now either too large or too dilapidated, or they do not correspond to modern environmental standards to be of any use. Indeed, many of these facilities will require expensive cleanup whether or not they are ever used again. As one cooperative manager lamented, "The worst thing is that we have nothing to show for these old debts. We built a cattle barn for 2 million in a building that did not belong to us. The *Treuhand* took back the building, removed what was inside, including food troughs, a milking installation, a tractor, and motors and trashed it. All that [equipment] cost money. The same occurred with the facilities for the workers, so that they could change and take a shower. The building is no longer utilized, but we are stuck with the 2 million." Wissing and Anz (1993:16) figure that 70.6% of the plants in which the fourteen Sachsen-Anhalt cooperatives in their sample had made investments were not being used in 1992. They contend (and our informants confirm) that "it is difficult to fulfill both demands (the servicing of the old credits and the payments of the legal claims of the LPG members) simultaneously and at the same time rationalize production by means of new credits. . . . Indeed . . . it is impossible without a temporary waiver to their claims on the part of both the farmers and the credit institutions" (1993:17).

In most instances, it would not have been to a cooperative's or its members' financial benefit to declare bankruptcy, for, as we have already indicated, liquidation would usually have meant incurring a greater loss than the amount of the debt. However, in some instances, the liquidation of an LPG and its subsequent reconstitution into one or more cooperatives would have been preferable. As the cooperative manager whose story we presented in chapter 2 conceded, "We should have liquidated the crop LPG and then divided up the land. . . . That way someone would have ended up with the debts but not the newly created cooperatives. But we were not astute enough to do that in 1990. Instead, we were well behaved and did everything in the right and proper. [And the banks] became creditors without ever having spent any money."

As of November 1993, it was still unclear to what extent the cooperatives would ultimately be held liable for the old debts. In October of that year the courts determined that such debts would have to be paid back in full. However, most of the cooperatives still hoped that a substantial part of these debts would eventually be pardoned. Nevertheless, as late as 1997, a legal ruling reaffirmed the validity of the debts. On the other hand, the *Treuhand* had already reduced the amounts owed by 25% in Mecklenburg-Vorpommern (Top Agrar Spezial, November 13, 1993, 3) a figure that was also given by our informants for their region. Most of our informants expected further reductions, perhaps up to 50%. Also, the amount a cooperative must allocate for interest and amortization of these loans is limited to 20% of its annual gross profit, which, at present, most cooperatives can avoid by taking higher first-year amortization rates or by distributing income in the form of higher wages rather than distributions. As one cooperative manager put it, "Mostly investments are made from the write-offs."

While the debts were still on the books in 1999, carried over from year to year on separate accounts with the interest, figured at the discount rate, compounded, offsetting the reduction in the debt instituted in 1993, our informants were no longer particularly worried that they would ever have to be repaid. As one informant put it, "In this respect, my mind is entirely free of worry" ("Da habe ich ein ganz sonniges Gemüt"). According to the same informant, a university was commissioned to do a study to determine whether the cooperatives would be financially capable of paying the debts back. There was also a government proposal to cancel much of the debt in return for lower subsidies on fuel. "All of that is really stupid because [the old debts] are going to be annulled in 2002 anyway. The Supreme Court has already ruled this. But how should Funke (the minister for food, agriculture, and forestry) know that? After all, he comes from the northern tip of Germany, where there is only water. He is a stressed human being."

A Case of Failure

With all the hurdles faced by the emerging cooperatives, it is not surprising that not all made it through the first years. Inexperience and mismanagement caused at least one collective in the area to go bankrupt. This occurred at the end of 1991, a little over a year after it was reincorporated as a smaller unit combining plant and animal production when three mixed livestock and crop farms were created out of two pure livestock and one pure crop farms. As one of our informants assessed the bankruptcy, "It need not have happened. It was due to faulty management and, let me put it that way, false friends. Those 'friends,' who were part of the management, owned large cattle trucks and within one year all the livestock had disappeared and with it a part of the assets. [Where did the money go?] For that, there is a saying, 'Nie sollst du mich befragen.' 'Don't ever ask me any questions.' The operation had assets worth 75 million and working capital of 3 million. And it had no debt. Imagine destroying all that! First, there was the decision to reevaluate the assets. For example, a tractor, produced in the GDR, that we had just acquired was reassessed at a book value of 1 German mark. We had purchased it for 40,000 [Eastern German] marks. Our irrigation system, which had cost 18 million (mentioned earlier), was valued at 0. In this way, the value simply disappeared. But it would not have been necessary. [To give you another example], two weeks ago I was in Bitterfeld. An acquaintance had telephoned and told me, 'Listen, there is a saddlery here and I am sure that you can use tent cloth for your trailers. Come and get whatever you want.' We came and loaded up and when we were done, there were still hundreds of rolls of tent cloth, leather, and things around and a bulldozer was already on the scene to load and move them to the garbage dump. You know how much all that is worth? Those are the problems with which we must come to terms.

"[To return to my original story], they would drive off to get such items as carrot seed. We had already purchased carrot seed from one of our producers. 'No, that isn't the right thing. We have to buy in the West.' So we drove to the West and bought carrot seed there. There, a kilo of carrot seed costs 160 to 180 deutsche marks. You require 9 kilos per hectare and we planted 180 hectares, so that was already more than 100,000 marks gone for seed. As the saying goes, 'A woman can carry more out of the house in her apron than a man can bring in with a wagon.' In the same way, a lot of people carry out more in their pockets than others can bring in. That is how it all started. There was nothing around any longer. Even the seed for brewing barley was simply no longer around. So they went out and bought and bought and bought. 'Others will pay.' All was fine. They signed checks subject to interest at rates of 6.8% to 7.5%. Then things began to fall apart. Who would pay now? The carrots were no longer needed. They had been sown, but no one wanted them. The peas were standing in the fields, but no one wanted them. The factories had been closed. Now what? Do you think you would find a single merchant who would say, 'God, they really cheated you, we will help you. . . . No, they wanted their money."

When the bankrupt cooperative was liquidated early in 1992, its assets, including a small factory, were purchased by seven local individuals—erstwhile members of the collective out of which the cooperative had been formed—who could thereby make a fresh start, unencumbered by debt. The manager of a neighboring cooperative explained: "Eight farmers had already joined to create a 400-hectare farm. They grabbed the 1,000 hectares. They are faring better than anyone else. They didn't have to pay the former members of the LPG, even though they themselves were once members of the same LPG. They didn't have the land before 1990, they started afresh. All they have to pay is the land rental fee."

Prospects

The statistics on agricultural cooperatives show a progressive stabilization of their numbers since the conversion of collectives was completed at the end of 1991. Their numbers continued to decrease between 1992 and 1998. That decrease slowed from 10.2% between 1992 and 1995 to 7.4% between 1995 and 1998 (Laschewski 1998:57, 193; Agrarbericht 1999). A similar stabilization can be observed in the average amount of land they cultivate: while cooperatives continue to lose land to individual enterprises, this loss has decreased over the years. Given the prevalence of plant closings in other economic sectors, "the stability of the successor enterprises to the LPG is quite spectacular" (Laschewski 1998:146).

These figures notwithstanding, the cooperatives, which consider themselves the democratic solution to agrarian reform after reunification, often still feel beleaguered by the new state and their other competitors. They were used to reacting quickly to new directives from above during the GDR and to dealing creatively with excesses in bureaucratic demands and the politics of agriculture. They thought that their rapid, flexible adjustment to Western rules and politics would be less frustrating and that once they had learned the rules of presentation and performance they would be accepted as legitimate actors in the new order. The new rules seem unclear partly because there are now many more avenues to success. What they are beginning to realize is that the struggle continues, and while the communist system kept all firms afloat, enterprises today do go under. They are aware of the seemingly arbitrary, often contradictory aspects of both hegemonic systems, but, while individuals could rise and fall in communism, enterprises were maintained; now, in the case of cooperatives, both individuals and enterprises are vulnerable. The cooperative mangers, however, acknowledge (as do the individual farmers) that state-favored units are similarly supported today.

The decision to maintain a cooperative form of agriculture in a society that frowns upon producer cooperatives and the constant struggle to secure access to land and to avoid the loss of land to independent farmers have profoundly marked the identities of cooperative farm managers. They regard themselves as providing the best chance for all but a handful of former members—those enterprising and educated enough and with larger than average personal holdings to enable them to strike out on their own—to continue in farming and to maintain a voice in running a farming operation. In a world of change, they regard themselves as representing continuity, but continuity that does not entail the obstinate clinging to an Old World order. Rather they see themselves as upholding an organizational form that was not allowed to come to full fruition under communism but that they regard as having tremendous potential to thrive under the new conditions if given the chance to develop without undue interference. They consider themselves as stewards of an old tradition dating back to the nineteenth century and as dynamic agents willing and capable of making the changes that will allow the tradition to be maintained under neo-liberal capitalism. The day-to-day decisions that inform cooperative operations to prevent failure and the manner in which the identity of the cooperative managers is affected by the struggle to cope under the new rules are the subject of the following chapter.

CHAPTER SIX

The New Cooperatives: Continuity and Change

The process that led to the development of the new cooperatives was characterized by radical ruptures, the combining of old and new organizational forms and practices, and, sometimes, the return to earlier ones. This experimentation with new forms and the recycling of old forms in new guises was evident in the restructuring of the manifold activities that the former collectives engaged in before the *Wende* and give testimony to the extraordinary flexibility of collective managers turned cooperative managers. That flexibility, but also the cost to rural workers in terms of unemployment, is manifested in the rapidity with which the east caught up with, and, in terms of labor productivity, even overtook the west as is evidenced in Table 6.1.

The decisions taken by the managers had to be made with an eye toward maintaining competitiveness in a context where profit margins are shrinking because of increased world competition and the reduction of state and EU subsidies. At the same time, one of the foundations of production cooperatives is the maintenance of workplaces for their membership. Worker identification with their work, fostered, among other factors, through the GDR policy of full employment, which meant that work outside the home was an important element in the lives of most working age individuals, further added to the seriousness of resorting to a reduction of the workforce to maintain profitability. The managers were thus placed in an unenviable position. A cooperative's very

TABLE 6.1
Labor and Land Productivity in Western and Eastern German Agriculture
(gross product in 1991 prices in deutsche marks, 1989–1994)

	Labor Productivity in GP per Worker			*Land Productivity in GP per ha*		
Year	*W. Ger.* *I*	*E. Ger.* *II*	*difference II vs I in %*	*W. Ger.* *IV*	*E. Ger.* *V*	*difference V vs IV in %*
1989	35,097	3,472	(911)	2,289	485	(371)
1990	38,224	8.146	(369)	2,413	798	(202)
1991	36,814	22,494	(64)	2,194	1,328	(65)
1992	47,812	43,175	(11)	2,720	1,467	(85)
1993	43,045	53,201	19	2,348	1,464	(60)
1994	45,863	55,852	18	2,357	1,354	(74)

Source: Reproduced from Thiele 1998:39.

survival was often predicated upon the undermining of one of its principal missions. The present chapter elucidates how cooperative managers navigated this dilemma.

Restructuring Activities

The decision to remain together rather than liquidate the LPG rarely meant the maintenance of the same activities or the same farm size. As we have already seen, the large, multifaceted units were considered by the western advisors and many of the farm managers as too unwieldy. On the other hand, they felt that the separation of certain activities and the specialization of farms into livestock- and crop-producing units mandated by the GDR authorities in the mid-1970s had been detrimental and needed to be reversed. Since, at the time of the separation of crop and livestock units, the latter were not consolidated to the same degree as the former, and each crop LPG served the needs of several livestock LPGs, the simplest way of accomplishing the reduction of farm size and of enabling livestock cooperatives to be more self-sufficient was to divide the crop LPG into as many parts as livestock LPGs it had furnished with fodder and allocate a part to each unit. But, as we shall see, this was not the only path taken.

Most important, however, many of the cooperatives' former ancillary functions had to be curtailed, for the ease of access to goods and specialized services that came with a capitalist economy made them obsolete. Specialized construction brigades, plumbers, carpenters, and electricians could be replaced by specialists hired only when needed. Perhaps just as significantly, mixed enterprises were now subject to higher corporate tax rates than purely agricultural enterprises and those whose earnings from other sources are limited, so it made sense to curtail subsidiary operations or to separate them out legally. Separation also made financial sense. After the *Wende*, wages in the construction industry rose much more rapidly than in agriculture. Such disparities within a firm could lead to dissatisfaction and to greater upward pressure on farm wages. Last, but not least, the creation of subsidiaries enabled agricultural cooperatives to limit adverse effects on the parent company, if a subsidiary failed.

As they were portrayed to us by the former manager of the farm's tractor pool, the efforts of a model state farm to privatize various parts of its enterprise stood out as an example of the more successful efforts to rationalize operations and, at the same time, conserve jobs: "The original idea was that [the entire repair department] would stay together. We knew that we could not subsist on engaging in repair work alone and so we wanted to build up commerce. But, after half a year, we saw that we could not continue with the large firm and so many persons. As a result, we rapidly came upon the solution of establishing a private firm. . . . [Earlier, we had already seen that we could not stay together with agricultural production]. The entire system was too cumbersome. I couldn't make rapid decisions. Before, ten persons had to be asked when one wanted to do something. It took a while to get used to the fact that one no longer had to ask permission from all sorts of people. I could now talk to a single person and then make a decision. Or I could decide on my own. In the past, one would first call a meeting to discuss the issue with five master craftsmen and engineers. After that, it would be placed before the next group, and then it would be discussed with the section heads, who would bring it in front of the director, and finally the funds would be released to buy a power drill for 100 marks. The system was extraordinarily cumbersome. . . . Everything had to be decided by general consensus. [The agricultural part] remained or rather became our client. We really solved this rather well."

The manager of this estate saw that things could not continue as before. He had social concerns: he did not want to fire individuals, and so he was happy when entrepreneurs came and hired workers from the enterprise. He explained, "We had to fire about half [of the workers]. But 'fire' in quotation marks. Our department, machinery, ultimately took over some ten workers out of the original fifty who used to be employed. A soft drink wholesaler that took over one of the buildings hired two or three more. Then, the filling station and repair shop were privatized, and they took over another ten or eleven persons. Then a metal workshop for construction work was established that took another six. I would say that a large number of workers out of our enterprise found work in the firms that emerged out of it. Finally, there was a large furniture store from Oldenburg that came over and hired another fifty persons. We had an excellent policy of attracting persons who wanted to do something, so that we could somehow accommodate our workers. . . . Our enterprise had been a model enterprise of the GDR. We had the means and a lot of recently constructed buildings, so many persons came over and could do something with them. I went to Mr. B to build something up or to make contacts, and the head of the former construction brigade also went to visit enterprises. Then there was a large cultural center where the furniture dealership is now located and they too went to look for a partner and, after originally thinking about establishing a fast food restaurant, ended up with the furniture store.

"But, in the last analysis, all of this was possible thanks to the far-sightedness of the management. . . . What remained of the enterprise, the construction department, the automobile pool, the metal construction department, and the restaurant still work together under the direction of the former manager. They constitute independent firms in terms of accounting, but they work together. . . . The director bought them and is working with them—some one hundred persons.[1] We ourselves became entirely independent in the summer of 1990, when it all started. Some six or seven persons have privatized or have become plant managers like me, or managers in other firms. . . . [We continue to do business with each other] because we know one another and also have fair prices. And, most important of all, when they have a problem on a Saturday or Sunday, then they come over to my house and say, 'I need this or that.' Then I come here with them and they can use the workshop, if they want to do something themselves, or they can take spare parts out of the store room. We do help one another."

Like the state farm in the example above, LPGs gave preference to firms that took over as many of their former workers as possible. For example, one LPG sold a farm building complex to an entrepreneur engaged in environmental activities, including tree trimming. A number of the LPG's former workers are now working for this firm. In other instances, craftsmen who had worked for the LPG were encouraged to open their own small repair shops and were even permitted to rent space in the former LPG buildings for that purpose, or use its specialized equipment.

However, these instances of successful transformation and continued cooperation should not obscure the fact that parts of many cooperatives had to be shut down completely. Gardening operations were particularly adversely affected. For example, one cooperative that continued to engage in many of its former activities nevertheless rid itself of the three gardening enterprises, only one of which was privatized by the former manager, a woman. She ran it as best as she could, buying whatever young plants she did not produce herself, and selling them in the market. The gardening enterprise of a second cooperative, where thirty to forty persons were once engaged in growing hothouse vegetables and flower seeds, was closed entirely; the hothouses became a vast complex of ruins.

Although the principle of streamlining operations while attempting to conserve relationships and jobs dominated the privatization process, the new cooperatives did retain some of the LPGs' multifaceted nature. Cooperatives continued to engage in some of the construction and repair activities they undertook before the *Wende*, even at the risk of disapproval on the part of the labor unions. As one manager commented, "I have had the union here once. Things have quieted down again. I feel that unions should not meddle in cooperatives. We can undertake certain jobs ourselves. No one can tell us not to do them. Of course, specialized tasks like changing around heating systems are done by outside firms. But we can do demolitions and adjunct tasks ourselves. . . . Among our seventy-five workers we have good people who also sometimes engage in such work part-time or for short periods. We have a few persons who are licensed as agricultural workers and who know how to drive tractors and engage in other agricultural tasks, but who also are professional masons. We take advantage of that fact. They lead small groups of two or three persons, where the others are not specialists or are simple journeymen." A second manager echoed this statement: "We are remodeling and renewing. People who can do these kinds of things in their own homes are engaging in these activities. They are upholstering, and ripping out floors. Of course, a specialized firm puts in the new floor." As we shall see later, some of the activities that are more closely related to agriculture, such as repair, were more likely to be kept.

Some cooperatives whose predecessors had combined agriculture with industrial production continued to do so. Thus, a group of eight farmers took over a factory that produces extracts from herbs that they grow on their own and on leased land.

The trend toward eliminating activities not directly related to agriculture, or, at best, maintaining some of the existing ones was balanced by a countertrend to expand existing services or even to add new activities. One cooperative not only did some of the farm vehicle repair work in house—something most cooperatives still seem to be doing—but did repair jobs for outsiders such as construction firms too.

Another cooperative, favored by its location close to a small city and its resources in the form of buildings, continued distribution directly to clients. It also produced crops such as onions and potatoes that everyone else had abandoned because of western German competition,[2] and it opted to keep its buildings as rental space. The company kitchen, where meals for workers of the collective were formerly cooked, was now used to prepare meals for schools and companies in the city (doubling production to 1,200 meals a day). A grocery store and a store for chicken, dog, and pigeon feed was housed in other spaces. In addition, this cooperative and two others transformed their fuel storage facilities into public fuel stations, and a fourth was planning to follow suit. One of these cooperatives modernized its gas station that lies on a major thoroughfare and added a modern car wash and a small convenience store. That same cooperative also ran a guest house with twenty-four beds. In 1994, unlike many competitors, it was very satisfied with occupancy rates.

Finally, some LPGs minimized the disintegration of the collective and the dispersal of assets, while embracing the flexibility and tax advantages of administratively autonomous units, by creating subsidiaries, fully or partially owned by the cooperative. One manager of a cooperative, whose astuteness, farsightedness, and organizational skills were admired by his colleagues and by independent eastern and western farmers alike, claimed that he was able to coordinate the various firms that had once belonged to a *Kooperation* during the GDR in much the same way that he had run the *Kooperation*, with only minor modifications in its actual operation. While he transformed the farm itself into a cooperative, he created a GmbH to purchase and administer land, sign lease agreements, coordinate work for other firms, and handle other administrative matters (presumably also external to the cooperative); a second one as a commercial enterprise; a third GmbH, as a construction company; and, finally, a butchery, which was actually an offshoot of one of the livestock cooperatives, as another separate unit. The cooperative owned the first and second companies outright and had a 60% interest in the third as well as an interest in the butchery. The remaining 40% of the third company was ceded to this branch company when the restructuring was undertaken. The managers sat in the same room and although the accounting was separate, they were fully integrated and were even allowed to offset profits from one unit with losses from another for tax purposes. Unlike many other cooperative managers, who limited their turnover from nonagricultural activities to 15% of the total gross income in order to avoid paying corporate taxes, the manager of this cooperative was of the opinion that there was no need to limit nonfarm activities. He argued that it made little difference whether the taxes were levied on corporations or on individuals, since the latter could deduct such taxes from the taxes imposed on the distributions they received.[3]

At the same time that cooperatives changed the mix of activities in which they were involved, they restructured the units engaging in actual agricultural production. As mentioned earlier, livestock and crop production were generally (but not invariably) rejoined. The joining of the two types of productive units involved a number of major considerations. Livestock-raising cooperatives were, in general, less lucrative than crop cooperatives, and many had incurred large debts during GDR times. On the other hand, livestock could be sold off for ready cash in hard times (even if at reduced prices), and the milk quotas established by the European Union could only be maintained when dairy farms were not liquidated. Although not specifically mentioned by any of our informants, there were also tax considerations. In Germany, livestock production that is dissociated from fodder production is considered a trade and as such is subject to higher tax rates (Hoell 1990:6).[4]

In addition, there were considerations of fairness. One cooperative manager explained: "In 1990, there were three farms that raised livestock in addition to the crop-raising unit. I divided—I must say that today I would act differently—the crop unit into three parts and allocated the land that they would have to lease to each of the livestock-raising units. I also assigned the machines, the people, the managers, and so forth to go with the new units. The state farm got all the land back that it had in 1971. We did all this because the livestock farm here was very weak. I could predict that it wouldn't last very long as it was. I felt that it would not be right for the people employed there who had originally contributed their assets. [If it had failed], they would not have received a mark, for the farm was also burdened with old debts [amounting to more than 1 million marks]. We had a few hundred thousand marks' debt on the crop farm, but I paid it right back. So we no longer had any debt. One of the livestock farms was a very good farm. At the time, we should really have joined the two livestock units and the two crop units into a single unit of 5,500 hectares. We would thereby have avoided a lot of frustration. But the good livestock unit did not agree to it because we had those old debts. As manager of crop production I might have been able to convince them to decide differently anyway. . . . [As things were], we were stuck with the debt, but also had the land."

Livestock and crop production were not invariably joined again. In one LPG, livestock and crop production went entirely separate ways. In a second, the cooperative that had established many separate daughter firms, two of the three collectives that had depended on the original crop collective for fodder became separate entities that raised fodder on their own rented land—500 and 200 hectares, respectively—while the third arranged for the crop cooperative to cultivate its land. In a third case, the crop and livestock units at first established separate cooperatives and then tried to unite the two forms of production. The manager explained: "While at the time of the *Wende* or shortly thereafter a large proportion of the cooperatives reorganized themselves the way they once had been with livestock and crop production forming a single enterprise, this did not happen like that here. I should say, unfortunately it did not happen. We missed our chance. At the time, there were two managements and they did not see eye to eye. In the beginning, the livestock farm wanted to become an independent enterprise with both animals and land, which would have left the other enterprise as a pure crop farm. But the head manager and some of the other managers were at an age when they could choose early retirement and they took advantage of the opportunity. So, as the situation had changed, we then said, 'Let's fuse the two enterprises into a single one.' But [the authorities] refused to give us permission. The reason was simple. The livestock farm with some 500 cows has a milk quota of 2.5 million liters of milk (per year). [The authorities] suddenly said that this quota could not be transferred to the new joint enterprise. Only Sachsen-Anhalt has this ruling. We had at least three meetings with the ministry, but, even after the farmers' association intervened, there was no way. They said that the quota had been transferred already once to the livestock cooperative and could not be transferred a second time. I can only conclude that the government of the *Land* has something against the formation of large enterprises. One of their arguments had indeed been that we were already a large crop operation and now we wanted to add a large livestock operation to it. . . . What has happened since then is that we (i.e., the crop cooperative) have not received any of the public land we once cultivated. Instead, 240 hectares were assigned to the livestock farm. With such a small amount of land, they can't establish an independent mixed farm. So, we have the following arrangement. Although we are not a *Kooperation*, in the traditional sense, the two units nevertheless work together. The livestock operation continues to receive fodder from us beyond what it can produce on the 240 hectares, at a price that does not permit us to make a profit, but allows their operation to make a profit from the value they add. . . . The advantage for us is simply that we can maintain the farm until we can unite the two farms. . . . [If we don't succeed], we are also thinking about giving the livestock farm the surfaces corresponding to the amount of land their members own, so they would have some 500 or 600 hectares. But we could also continue with the present system."

In this example, we find the ironic situation where the western-imposed legal system prevents two cooperatives from instituting a system that the experts have themselves promoted: the joining of crop and livestock production. Instead, the ruling led to the reinstitution of the 1980s system of cooperation between legally independent units. It should be noted that the western experts' advice did not correspond to the general practice in western Germany either, where there was considerable specialization in crop and livestock farming, the main difference with the GDR separation being that livestock farms usually had the land to grow a large part of the fodder consumed by the animals they raise.

One cooperative in our sample, the same one that continued to engage in activities outside agriculture in the form of GmbHs, retained the separation between crop and livestock production; but the units worked in close collaboration. It appeared that in this case the *Kooperation*, the GDR organism in charge of the coordination between plant and animal LPGs in a given area, had always been stronger than it was in other localities and so the actual managerial structure remained the same after the *Wende* as it was before. Animal production and plant production were conducted independently, but economically the two units appeared to be integrated, like in their other GmbHs in which they held an interest.

In 1993–94, the organizational forms and practices in eastern German cooperatives, then, entailed both a variety of continuities with the past as well as the resumption of intermediate patterns and a partial, but not wholesale, acceptance of western-derived and newly devised models themselves often instituted by western experts. By the time of our second visit in 1999, further shifts had occurred, mostly in the direction of simplifying the structure of the enterprises and of making additional improvements and investments in its most successful activities. Two of the cooperatives we revisited made radical changes. The first closed its dairy operations. In order not to lose its milk quota, it made an arrangement with a cooperative farther north to take over the herd in return for a share of the profits. In the north, land is poorer and less suitable for growing crops and therefore cheaper to rent, and so both cooperatives could benefit. The second, the cooperative that had created subsidiaries in which it held shares, lost most of these subsidiaries altogether. The two livestock units and the construction company in which it had held interests[5] all went bankrupt. The interest in the meat-processing plant was sold and its resort facilities liquidated. These changes did not affect the cooperative's financial health. Indeed, it was the first to pay off its debts to former members and it was able to make a 1.5 million deutsche mark investment in an onion storage facility and to purchase the largest amount of land—180 hectares—all without incurring loans. It remained the most successful cooperative in our sample. The shedding of these subsidiaries does not mean that the cooperative has abandoned the concept altogether. It still runs a commercialization enterprise and administers the land it cultivates that is owned by unknown persons in this manner. The latter enterprise was originally created in order to attempt to establish a second farm, a venture that was thwarted by the tight market for land. The present manager of the cooperative explained the rationale for subsidiaries as follows: "We always thought that it was better to create independent subsidiaries so that the parent firm would not be affected."

The cooperative run by Mr. Werner, whose case we described in chapter 2, closed both its hotel and its gas station/car wash operations. With the decrease in the number of westerners who engaged in construction projects in the east and the fact that workers working on the nearby throughway lived in barracks rather than hotels, there were too few guests to make it worthwhile; also the *Land* was not willing or not in the position to sell them the facility. The gas station would have required an expensive replacement of its tanks and was located close to other stations. On the other hand, they have signed a thirty-year contract with a solar energy company to install 16 windmills on land they cultivate, for each of which they receive 1,000 DM per year for the inconvenience.[6] When we revisited Mr. Werner, he was in a negative mood. He had just read about the government's plans to further reduce subsidies and price supports for agricultural products and was worried what that would mean in terms of further reducing personnel and perhaps even abandoning money-losing but work-providing activities such as pig raising. In spite of improved yields and good harvests, the cooperative's income had decreased by 30%, barely covering expenses. Mr. Werner himself was planning to retire soon and hand over the reins to a 30-year-old woman whom he had trained himself.

However, not all of the cooperatives we revisited got rid of subsidiary operations. Another cooperative that ran a gas station continues to do so, even though it meant replacing the tank to conform to western German standards. As it turned out, the old tank had not leaked at all, as the present manager pointed out disgustedly; but rules are rules. And yet another cooperative that had continued to engage in many additional activities also did not give them up. Finally, the cooperative that also ran a plant extract factory, still produced plant oils but was forced to reduce its operation and cut its workforce in half to three employees.

FIGURE 6.1. Modernized gas station of a cooperative.

In conclusion, the strategy of maintaining employment by continuing a high degree of diversification, while not abandoned entirely, has fallen prey, at least to some extent, to cost-benefit calculations. When diversification is upheld, it is sometimes justified if it contributes to lowering costs in the agricultural operations (e.g., the gas pump), or if it does not involve significant additional labor costs (e.g., the windmills). In eastern Germany, then, one inconsistent system was replaced by another. Communist firms were characterized by specialization to increase volume, but counterbalanced by the need to become as self-sufficient as possible to remedy problems of input allocation. After the *Wende*, the cooperatives attempted to both streamline operations by outsourcing some of their activities associated with farming, but also to retain some of their former diversification to curtail unemployment. But the argument of self-sufficiency as economically desirable has not been entirely abandoned either. Paralleling the global bifurcation of tendencies to outsource and countertendencies to merge activities, the specific economic history of such antagonistic trends continues to lend the development in eastern German cooperatives a distinctive tension.

Investment in Buildings and Machinery

A visitor to a cooperative is struck by the simplicity of the buildings. Some of the barns predate the GDR. They were once part of large estates or demesnes. Many of these are falling apart. Some were abandoned in GDR times, others destroyed by arsonists after the *Wende*. Those constructed during the GDR are functional, but rarely well built. Office buildings are often barrack-like. Even by 1999, managers had invested little, if anything in office furnishings, so the visitor is ushered into surroundings of a bygone age, stark but not unfriendly: an old desk or two, a table for staff meetings, chairs, and little else. There is no correlation between the financial health of the enterprise and the image it presents. The offices of the most prosperous cooperatives were located in one of the least attractive building complexes and, although the cooperative had invested heavily in onion-processing equipment, the millions of deutsche marks invested didn't show from the outside. Cooperatives have indeed had to make major investments in technology in order to bring their operations up to date. Milking equipment was replaced, barns were transformed to make them more efficient, and storage facilities improved. Most important, the old and outdated machinery pool had to be renewed. Financing the purchase of tractors and other farm equipment either through bank loans or dealer loans was relatively easy,

FIGURE 6.2. GDR and modern western farm machinery and old and new barns of a cooperative.

as the machines themselves could be used as collateral, but required a down payment of 25% to 30% of the purchase price. Machines were also acquired by means of lease-purchase arrangements. However, cooperatives suffered from insufficient access to bank credit for construction, installations, and even land purchases. Banks appeared to be reluctant to lend to cooperatives because of the uncertainties in the latter's access to land and unclear title to the land on which farm buildings stood; lack of personal connections between bank and farm managers; and, in many instances, the large debts incurred by the cooperatives during communist times. A further problem was the fact that cooperatives had an incentive to underreport their assets in order to reduce the claims to compensation by former members. This also made them appear as less credit-worthy to banks (see Hartmann 1995:113–114).

While cooperatives, like their private counterparts, benefited from a variety of subsidies including low-interest loans, they were excluded from public loans and start-up grants conceded by the government to private farmers with annual family incomes under DM 100,000.[7] However, they did receive substantial adjustment payments right after the *Wende* to permit them to slow down the reduction of the workforce.[8]

Some cooperatives made substantial investments out of their own income.[9] Thus one cooperative manager estimated that they had invested between 3.5 and 4 million deutsche marks in machinery and farm buildings, and an additional three-quarters of a million in a construction firm in which the cooperative had a financial interest. He claimed, "We don't have any credit. We paid cash for everything. We only bought a few tractors on an installment plan of three installments of 40,000 marks each. We have paid it all off. I don't believe in credit. One should give credit to others, one should not request it."

Strategies concerning technology ranged from rapid replacement to more conservative ones. The cooperative managers used their knowledge of the once difficult to obtain tractors from other former COMECON countries to reduce costs. They were also happy to finally be able to acquire equipment with the power that was commensurate to the tasks at hand rather making do with underpowered GDR-built tractors. In addition, many types of specialized labor-saving equipment, which were simply not available in GDR times, necessitating the continued reliance of hand labor, were now easily accessible. Indeed, U.S.-manufactured John Deere tractors and other makes adapted to large-scale farming and considered relatively maintenance-free, became eastern German favorites; but even the low-tech and cumbersome, but cheap and powerful Russian tractors, which were now readily available, remained popular.

Cooperatives varied in the way they bought and maintained their equipment. Some cooperatives had replaced most or all their tractors and combines by 1994, shopping around for the best prices among the numerous suppliers from the west.[10] As one cooperative manager explained, "There are many dealers, which is reflected in the prices, and so we bought two combines from the John Deere [dealer] and four tractors from two other dealers. When one asks, give me your price and he tells you the price, then you buy where it is cheaper. The service is pretty much the same, and so we have established the base with 115 horsepower tractors and 210 horsepower combines. In addition, we have purchased a new German harvester, a Deutz, with a 280 horsepower engine. Our largest tractors have 170 horsepower engines. More powerful ones are unnecessary and merely cost more money. Three combines serve us well to harvest 1,000 hectares. Last year we still used two old combines produced in the GDR and they did quite a good job, but now they have given up the ghost, so they have to go. Their cutters are used up and so forth. . . . We do have to keep everything up to date. But we have to be careful about what to replace each year. We don't have much money at our disposal."

A second manager claimed to have replaced half of the cooperative's mobile equipment in the preceding four years, arguing that even relatively new eastern German combines were not as effective as the western ones and that farms that continued to operate with eastern German technology alone had produced unsatisfactory results.

A third cooperative manager preferred to maintain the old equipment as long as possible and was also more willing to purchase Russian or Czech tractors. "One can't get a western tractor for less than 80,000 marks for a 100 horsepower model, while a similar Russian or Czech one costs between 30,000 and 35,000. . . . The quality is not as good. The finishing is inferior, but not so bad that the machines are only half as good." However, this manager, who had words of praise for the eastern German farm equipment, also purchased John Deere machines. Interestingly, he justified this choice in part on the basis of that company's long-term association with Germany. "For us," he explained, "this brand is of interest because the firm has its origin in the fusion with the German firm Lanz. The Lanz Bullfrog was one of the first tractors, with only one piston. That firm made many good farm machines. The John Deere machines have the advantage that they are the machines that are best adapted to conditions on a farm. They are not too complex; they do not have too much electronic circuitry, which can and does go wrong. Wear and tear is relatively low. We have two combines from this firm and a tractor that is already in its third year. In those three years we have had very few repairs, even though the machine is under heavy use on our 1,500-hectare farm.[11] We are going to continue to give preference to this firm."

The argument for maintaining old equipment for as long as possible was made most forcefully by a cooperative manager, who linked the strategy to the loss of land experienced by the cooperative: "We could have [invested in new machines]. We didn't though, because we reduced the number of animals. The surface we cultivate has also been reduced: a few farmers have started their own private farms again and old landowners have received their land again, so that out of 3,500 hectares, we have lost 1,200 between 1991 and 1993. Every year we lost some 400 hectares, but the machines of course remained. Often, they were old machines, but we didn't just trash them. We follow the principle 'out of three make two,' and use the machines until they give up. We have repair facilities with fifteen workers and repair everything ourselves. That is relatively cheap. The external repair shops charge 35 to 50 marks per hour, while it costs us only 20 marks: 15 for wages and 5 for other expenses. We also have new machines, for example, three combines and three new Massey-Ferguson tractors. We have new plows, new ridge-drills, and new manure spreaders. But whatever is still serviceable is driven until it stops. Then it is still not thrown out. Instead, we make two out of three. One doesn't get anything if one tries to sell them. Many farms have sold old 90 or 100 horsepower tractors for 1,000 marks. . . . When I have thirty such machines, I am not going to sell them for 1,000 marks when I know that the tires alone cost 1,500. So they sit there and when the treads are gone on one set, we exchange them and the tractor runs for at least another year. The same goes for the driver's seat and so forth. There are lots of parts, like headlights and things like that, which can be reused. Ultimately, that's worth much more than 1,000 marks. I figure a value of 4,000 to 6,000 marks, an amount that no one would pay for an old machine."

Still another manager followed the same strategy, arguing that "We don't need the newest machines, our old tractor will do. A new machine is gone after ten years while land may be worth double." This manager was least enthusiastic about western technology. He took us for a drive to show us the cooperative in his GDR production Wartburg and complained, "We bought a Mercedes tractor for a lot of money, because it's supposed to be something really good. But the gears have broken down already, even though we have a good tractor driver driving it. Now we buy 80 horsepower Russian tractors again, like we always did. Each one costs 25,000 marks and we are familiar with them. They run ten years too. Some of their features are primitive but [the tractors are] functional. We also want to buy a 150 horsepower tractor from the Czechs for 50,000 to 60,000 marks."

Cooperatives often made use of the excess capacity in their farm equipment pool by doing work for private farmers or, in one case, for a neighboring livestock cooperative that decided to venture on its own.

Ironically, many of the underpowered old tractors, which were frequently used by collectives because they were produced in the GDR and therefore were more readily accessible, were finding their way into small western German farms when they were not shipped off to the Middle East.

The strategies of the cooperative managers concerning equipment was, then, again a mix of conserving the old and experimenting with the new. Overall, these strategies certainly contributed to lower labor/technology ratios but in ways that were more subtle than one might expect. GDR agriculture had long been highly mechanized but it included labor-intensive components. Weeding was often done by hand. New equipment combined with more effective herbicides obviated such practices. And, larger and more powerful tractors and combines could do the work faster and with fewer operators. The effect of the latter factor had become particularly pronounced by 1999, when cooperatives had replaced much of the equipment they had purchased during the GDR. The new combines, for example, could do the work of at least two old ones. As one cooperative manager commented, "We still used to have four E16 (eastern manufacture) combines. We have a new one and got rid of all four of the old ones. The new one doesn't do the work of the four old ones but certainly that of two." Perhaps most important, previously, the difficulty of replacing equipment demanded a large labor force to maintain it well beyond its usual useful life. That is no longer necessary. In addition, the power of tractors can now be more readily be calibrated to the task, again saving labor.

Finally, none of the cooperatives we visited are employing satellite-directed technology on their farms. One informant explained, "This is a satellite photo of the area. My son has written his thesis about global positioning systems. I believe that it is not yet feasible to employ it for everything, because the technology is still too expensive. One is supposed to be able to spread fertilizer in a modified manner. But you can't know what the rain conditions are going to be like. If I have 900 to 1,000 millimeters of rain, then I can do it. But with an average of 450 millimeters it is more difficult. The nutrients are not always the deciding factor; rather it is the water that comes from above. Also, at this point, one has to spend a lot of money for it. A combine costs 20,000 marks more. For six combines that already makes 120,000. Then you have to install sprayers and fertilizer distributors. It is questionable whether it makes financial sense. For small farms, it doesn't make sense anyway. They are trying to open up a market niche, but the question remains whether it is a good thing for the farmer or for the dealer. I believe that it is good for the latter." A long history of technological innovation and knowledge of local growing conditions, then, allows our informants to make carefully calculated purchases.

Crop Production

Although, in 1994, the cooperatives studied had not entirely weathered the turmoil in agricultural production that followed monetary union and reunification, the elimination of entire spheres of production had already largely run its course. Cooperatives continued to engage in crop rotation, although less strictly than before the *Wende*. Blessed with soils of extraordinary fertility, often easily tilled loess, the region is well known for the high quality of the wheat it produces, for which the relatively dry climate in the rain shadow of the Harz Mountains is an asset.[12] As one cooperative manager put it, "If you don't forget to sow, you can count on a good yield." Since cooperatives can offer large lots of grain to intermediaries, they can also command relatively high prices for a given grain quality.

In 1999, one cooperative manager described the proportions of land typically dedicated to different types of crops—much of which is determined by the amounts subsidized by the state—as follows: sugar beets 8%, rape seed 6%, wheat and barley 70–75% (he plants 4 times as much wheat as barley), fallow or planted in crops for fuel 10%, and the rest in peas (for feed) and small percentages in other crops.

By 1994, rape seed, used to produce cooking oil, and, on an experimental basis, fuel, had grown in importance. The more adventurous farmers had also begun to grow sunflowers for cooking oil, a plant not usually grown in the region.

Sugar beets continued to be a mainstay of the local economy. The production of sugar and, as we shall see later, milk, are subject to quotas which, in turn, at least in the case of sugar beets, ensure adequate profits. In western Germany, the state governments in conjunction with the sugar industry originally established these quotas in the early 1970s according to the average planted in the six-year period between 1966 and 1972. In the mid-1970s an additional allocation was made. Thus, farmers in the area of Hannover and in southern Germany are allowed to plant 30% or more of their land in sugar beets. In contrast, in 1994, farmers in the new *Bundesländer* were only allowed to plant 10% or less of their land in sugar beets, even though some LPGs had dedicated as much as 25% of their land to that crop. With increases in productivity, the area that could be planted in sugar beets had to be reduced commensurately. Thus, while in 1990, 201,000 hectares were still cultivated in sugar beets, that figure had shrunk to 146,000 by 1994.

By 1999, there was a further reduction. According to one western German informant, this disparity was a major reason for the fact that land rental prices were much lower in eastern than in western Germany. Thiele (1998:178–180) concurs with this informant's claim that the eastern German *Länder* were treated unfavorably by the authorities. The sugar quota is 21% lower in the east. And for grains and oil seed it was already evident through the overproduction in 1993 in the east, but not in the west, during the first year the regulations were applied, that the quota was too low.

The cultivation of hops, an important crop before the *Wende*, was abandoned by some cooperatives, even though they had elaborate trellis complexes made of cement pillars and wire. They also had drying installations. One cooperative decided to bulldoze facilities that had once cost them 1.5 million eastmarks to build. Another cooperative that owned drying facilities completed just before the *Wende* continued to grow hops on 25 hectares of land, but only because the facilities required no new capital investment. Nevertheless, a third cooperative, where four persons were permanently engaged in tending hops plantations, another four during the growing season, and more than thirteen during the harvest appeared to be in the business for the long haul. It had introduced a North American variety, reputed to be particularly well adapted to a dry climate, and plans to slowly replace the existing varieties, for which the demand is low, with new varieties.[13]

Except for onions, which two of our cooperatives were still producing and marketing successfully, none of the cooperatives in our sample continued or reinstituted growing fresh vegetables. They felt that they could not compete with other countries with more experience in producing high-quality produce, better climatic conditions, and fewer restrictions on the use of fertilizers and pesticides. One cooperative manager commented, "Last year we had a colleague from the agricultural school of B. She was here for eight weeks and worked along with us in administration and with the combine. She did a bit of everything to gain insight and is now finishing her studies. When the combines were working until late at night, we would sit together and talk. That's why I say, we can't compete with them, the Dutch are not subject to any limits in terms of the amount of fertilizer and pesticides they are allowed to use. We are limited and can't follow suit. The Dutch can achieve enormous yields on very small surfaces. Our use of fertilizer is also limited by considerations of cost, but we are interested in ecology as well as economy. They can put downward pressure on prices without hurting. We, on the other hand, are also limited because of the amount of land we must fallow. After all, we have all studied agriculture. We want to be able to utilize the land for many years, not just five. For that reason we do not use liquid manure; we work with straw and farmyard manure, which we plow into the fields. If I did not use straw, I could have sixty or eighty cows for each milker, while with straw, which has to be spread in the barn, I can only have forty to forty-five cows, raising labor costs. But I don't like liquid manure. I am not allowed to spread liquid manure on the land between October and February. So I would need enormous tanks to store the liquid manure for five months. That would cost a lot of money too. . . . For us, the cost per cow space is 10,000 marks. If we switched to liquid manure, it would cost us 15,000. In general, we do not engage in plowless farming. We only use the latter method in fields previously planted in sugar beets where we plant wheat. . . . But we can't compete with Dutch vegetables. A substantial amount of knowledge and experience is required that the Dutch have achieved by growing vegetables for many years, an experience we can't match."

Similarly, the future of fruit production remained in limbo, and those producers who had not already cleared away their orchards were uncertain whether they might not have to follow suit even with orchards that were about to reach peak productivity.

The region is well known for its medicinal and culinary herbs and spices, which some cooperatives were continuing to cultivate. For example, one cooperative cultivated 100 hectares of marjoram, peppermint, digitalis, balm-mint, and other herbs. Another, small cooperative, that extracts herbal oils, planted 40 hectares of marjoram, 10 of parsley, 6 of celery, and 1 each of tarragon and oregano (organum), as well as dill and dear stick[14] (*Levisticum officinale*). Still another cooperative planned to grow camomile and was gearing up to produce ergot[15] in large quantities.

A particularly enterprising cooperative manager, who adapted to the new situation so rapidly that the cooperative did not experience a period of crisis after the *Wende* at all, expressed his attitude toward production as follows: "I have a principle. If silk stockings bring money, I cultivate silk stockings." Unfortunately, as we shall see later, these unusual and/or innovative strategies did not necessarily translate into higher profit rates.

By 1999, several changes had occurred in crop production. Most important, prices for farm products had fallen substantially, reflecting the general decline in world prices for farm commodities. For example, first-quality wheat sold for 23 deutsche marks per 100 kilograms, a drop of 5 to 6 deutsche marks from a few years earlier. Peas sold for 20.50 deutsche marks per 100 kilograms, a drop of 9.50. The sugar price remained the same for the assigned quota, but anything beyond that fetched a much lower price than before. In addition, the sugar factory told the farmers to reduce the acreage devoted to sugar beets by 5% in 2000. One cooperative manger lamented, "We had very good harvests in the past three years with very good efficiency, but if one compares last year with the very bad year we had three years ago, there is very little difference [in profit]."

Interestingly, it no longer seemed to pay to produce top-quality wheat. Wheat of an average quality paid better per hectare. As one cooperative manager said disparagingly, "The mills prefer B and C wheat to make flour. The bakers don't seem to care. They only add 10% extra-high-quality flour for their buns which should really be made out of the best flour. Well, you don't need a baker to bake that type of bread. A mechanic could do that too."

There were few innovations regarding crops. Indeed, such new crops as sunflowers were dropped again because of poor results in the region. When linseed prices were highly subsidized for a year or two, growers experimented with that crop, but did not like the fact that the tough plants were hard on equipment and they ceased growing it when the subsidies were lowered again. And the cooperative that had still maintained large apple orchards in 1994 uprooted the trees on 15 out of the 39 hectares in the past five years, for only direct sales to consumers still made economic sense. They intended to take advantage of a proposed subsidized campaign of the European Union to reduce fruit trees and remove another 10 to 12 hectares but to plant 2 hectares of new varieties. On the other hand, the cooperative that introduced ergot in its crop mix continued to expand its production. The cooperative that also used to produce plant extracts had added Saint-John's-wort (hypericum) and Unica Specia. But the demand for extracts of these plants soon declined again. As one cooperative manager put it, "The experimentation is taking place in the types of pesticides and fertilizers we employ and in new ways of sowing, for example, plowless cultivation, rather than in the cultivation of new crops, because the market is limited. What is the point of growing them when one can't get rid of them? There are limits." One of these limits is certainly the fact that genetically altered crops are not permitted in the European Union. One informant commented, "The fact that people are working on these matters must be evaluated positively. They talk about it too much in the media. It will be the only way to deal with the explosion in the world population in the next ten to twenty years. Yet, I am very much for strict controls of beef from England. That should be even stricter. But genetically altered corn or tomatoes? That would be nice. For example, there are so many tasteless tomatoes on the market. [There is a lot of room for improvement.] That is my personal opinion."

In conclusion, the rationale for choosing to plant a particular crop has, on the surface, changed dramatically. In the GDR, the choice of a particular crop was dictated from above. However, such choices could, to a degree, be influenced at the local level, presumably mainly by adducing historical precedent, but also if a manager had connections with the authorities in charge. While demanding, the cultivation of crops that were out of the ordinary could provide a collective with a shelter from close official scrutiny and enable it, for example, to keep its own planting schedule, pay its workers a higher than average wage, or even allow its members to continue individual livestock raising. After the *Wende*, such decisions were ostensibly dictated by the open market. That market, however, is, in fact, strongly regulated by the EU and, as we have seen in the introduction to Part II and will explicate more fully later, by the practices of monopolistic industries and the well-established network linkages of western distributors. Many of these market-driven changes, both in the types of crops grown and in the manner of cultivation, are geared toward saving labor and thus contribute to rural unemployment.

Livestock Production

By 1994, the precipitous decline of livestock that occurred in the first two years after unification had ended. Cattle herds had stabilized at around half of their original size or were beginning to increase again, while pork production was still declining. For example, one cooperative that once raised 950 cows had stabilized the size of its herd at 400 cows, a second at 550 (from its original herd of 980), a third at 600 (from its original 1,000). Another, which had reduced its herd of milch cows from 720 to 250, planned to increase that number to 300 by the end of 1994.[16] The same cooperative had originally planned to follow the example of many others and get rid of beef production, but decided to acquire more bullocks for fattening instead because of the low cost of sugar beet residues due to the bumper crop in 1993 processed by a new sugar refinery.

The cooperatives were able to almost double the amount of milk produced per cow thanks to concentrated feed that was now readily available, and by optimizing herd size. The animals they sold were a welcome cushion during the turbulent first years and the drought year of

1992.[17] At the same time, in the mid-1990s they were still struggling to reach the hygiene standards mandated by the Ministry of Agriculture. In 1994, these standards were raised further, which was expected to lead to a scramble to make improvements and, perhaps, for many farmers to abandon milk production altogether (Altmann 1993). As one cooperative manager explained, "[Before the *Wende*] we were not allowed to choose between keeping 550 cows and producing 4,000 liters per cow per year or keeping 700 cows and producing 3,000. The whole world knew that we were too stupid to produce. Now we produce 5,500 to 6,000 liters with the same cows, but with better people because the worse ones are no longer working in the barns. The basic amount of feed was simply too low. . . . We always achieved the highest production from May to mid-June. Now we no longer have such a jump in production when we bring the cows out to pasture. We produce 1,000 liters more in total. Now we not only feed the cows silage all year round and add the green feed, but also provide more feed in general.

"We produce Category I milk, but we have also produced Category II milk. We don't achieve the top rating every month. We still have problems with the cell count in the milk. We have a new milking installation equipped with refrigeration, where people practically no longer come into contact with the milk. They also handle the udders very skillfully. But it is difficult to influence the cell count and as soon as it reaches 400,000, we lose the Category I rating. That means a loss of 0.02 per liter. We are happy when we achieve the top rating and also proud, but we don't always succeed. It's too bad. It depends on the age of the cow. Also, with the old milking machine we ruined many an udder because we had to depend on the milkers to prime the udders. Because we needed many milkers, their qualifications were not always what they should have been. [That explains why] the older cows produce milk with a high cell count because they slough off a lot of cells. Now it's different. One would think that the younger cows would be okay, but that is not quite the case either. Perhaps it's genetic, but I don't think so. There must be other factors. A cow reacts to unusual stimuli, such as loud noises and even the mood of the milker. When the controllers of the *Land* come on a day when a milker was in a bad mood and pulls a sample from the tank and examines it, one suddenly has a count of 450,000 rather than 300,000. . . .

"Every year we add 20% to 25% new cows through reproduction. Last year we again eliminated cows with an early stage of Leukemia (*Leukose*). Although we no longer have any cows with *Leukose*, the herd has to be free of *Leukose* for two years before this is officially recognized. . . . As long as animals are isolated and the *Leukose* does not

FIGURE 6.3. While many of the farm buildings of the cooperatives are still old, the dairy herds are up to modern standards.

FIGURE 6.4. Some cooperatives still prefer to use straw as bedding for cattle in spite of higher labor requirements.

advance to Leukemia, one can live with it, but the entire old *Bundesländer* are free of the disease and the new *Bundesländer* must now also achieve that status. So we got rid of all the affected animals and replaced them with new cows that had already given birth. We have already passed four controls, and face an additional three."

Dairy producers were not free to increase production at will. Milk production, which was heavily subsidized by the European Union, was subject to stringent quotas that, in the case of eastern Germany are based on production in 1989 and, unlike in western Germany, are not transferable. The rules stipulated that farmers could produce a maximum of around 80% of what they produced in 1989,[18] for which they received a one-time compensation of 3% of the value of the milk produced in 1989 and an additional annual compensation that in 1991–92 came to about 1.5% of the same amount. While dairy producers in Sachsen-Anhalt had generally not yet reached their quota limits by 1994, they had to factor them into their longer-term planning. The milk quota per hectare in the old *Länder* is almost twice that in the east. In addition to higher quotas in the west, farmers there were allowed to buy milk quotas from other farms in a given district. Cooperative managers hoped they would be able to do so in the east as well, providing them with greater flexibility. In the meantime, the only way they could make use of their milk quota if they no longer wished to continue to raise cows was to engage in the contorted transactions of the kind employed by the cooperative described earlier, which made a deal with a farm in another region.

The greatest worry of the manager quoted above was that if the cooperative should lose the *Treuhand* land it now cultivated, it might have to stop dairy production, or, at the very least, it would have less leeway to weather a drop in the milk quality ratings. It would also have to get rid of its labor-intensive pig-raising operation. As things were, it planned to rationalize the operation. It was still raising some 300 pigs. These were dispersed in a number of barns holding sixty to a hundred each. As the manager complained, "For sixty pigs one person is really too much, while one person can't handle a hundred animals. Such factors influence our personnel costs. Consequently, we will reduce pig raising to a single building to lower personnel costs and to streamline operations. This will also enable us to save on rent, since we are still renting one of the barns, which is making pig raising unprofitable with the prices as they are at present and are likely to remain in the future." The plan was to keep 150 pigs in the remaining barn tended by two farm hands. Pig raising also had to be adjusted to western notions of meat quality. In 1990, eastern German farmers were considered to be hopelessly behind their western counterparts in terms of fat content of the meat they produced. By 1994, however, our butcher informants were becoming alarmed by the watery and tasteless meat that resulted from the efforts to reduce fat content and some of them were selecting carcasses that were closer to earlier norms (see Grund 1994).

Earlier on, the same cooperative also got rid of its geese, which it raised in the open. That operation was no longer lucrative enough and the 10,000 geese hardened the ground with their feet and increased the danger of salmonella disease. As we shall see later, the cooperative's sheep were taken over by the former shepherd, who continued to keep 300 ewes.

The drastic reduction or elimination of sheep herding and the reduction of cattle and pig holdings had resulted in a large amount of unutilized barn capacity. On the other hand, according to Top Agrar Spezial, one-fifth of the 78,000 cattle and pig barns in the new *Bundesländer* could not be used profitably in a market economy (November 15, 1991, 13). Indeed, the landscape was littered not only with the ruins of barns dating from precommunist times that fell into disrepair under communism, but of newer buildings as well.

Many of the time-consuming activities fell by the wayside. Thus, a cooperative might no longer make silage out of beet leaves and might cut up and plow under the straw rather than bringing it into the barn. Also, whereas formerly LPGs would transport sugar beets to the factory themselves, now the factory picked the beets up itself.

In sum, with the removal of state dictates requiring the maintenance of herds beyond the capacity of the land, the availability of high protein feed additives, and a massive reduction of the workforce, the cooperatives showed a vast enhancement of productivity per worker. This improvement occurred in spite of the drop in demand, the need for cash to continue operations after the *Wende,* the uncertainty of the future of collective and state farms, and, in 1992, one of the worst drought years in decades, which led to a massive sell-off of animals, often at ridiculously low prices (gross production was reduced by 36% between 1989 and 1992 [Schmidt 1992:14]).

Whether the extraordinary increases in productivity experienced by the cooperatives are sufficient to offset sinking milk and pork prices and rising land rents in the fertile region of Bernburg remained an open question.

In 1999, none of the cooperatives we revisited seemed to be making any profit from their livestock operations; indeed, these were subsidized by their crop operations, all this in spite of the fact that most of the technical problems such as high cell counts in the milk had now been solved. As we have seen earlier, we heard about the failure of three pure livestock cooperatives. The cooperatives were loathe to stop raising animals, for such a radical move would greatly increase unemployment. As we have already seen, only one cooperative did so. In that cooperative, dairy production had been in the red for several years. It had made relatively low investments to rationalize livestock production and was therefore in a good position to walk away from dairy farming. The manager invoked history to justify the move. He claimed the region had traditionally concentrated on growing crops. "Before World War II, the landed estates raised cattle mainly in order to remunerate their workers. The estates would raise twenty cows or so. The milk was destined for the consumption of the workers. This was the so-called *Deputatenmilch* and butter. As late as 1945, my parents still received milk and butter in addition to cash. The cattle were fed sugar beet residues. Now the factories extract more of the sugar, but then some 5% remained in the bagasse." The decision to stop raising cows has increased profitability very considerably. It also was the main factor in the reduction of the workforce from sixty to twenty permanent and ten seasonal workers. The same cooperative also changed its pig production. It has decreased the number of pigs it raises from 2,000 to 1,500, saving the labor of one worker. But, rather than keep the pigs until they reach the usual weight of 90 to 100 kilograms and then sell them to large-scale meat-processing plants, they fatten them until they reach 180 to 200 kilograms and then sell them to small-scale butchers, who prefer the fatter pigs for the so-called *Hausschlachte* or home butchery.[19] Also, to make good sausages, the meat has to be processed when it is still warm. It should never be frozen. As the manager commented, "I am amazed at the way in which the population increasingly demands this type of home butchery product." The cooperative also made major steps at rationalizing the operation. Labor could be saved by changing the configuration of the barns and even the materials of the dividing walls. As the manager explained, "Six years ago we still raised most of the pigs in primitive facilities built during the GDR with straw bedding rather than liquid manure. Four or five shared a box. The boxes had to be cleaned daily and the pigs had to be fed, in part, by hand because the passageways were too narrow for a vehicle to navigate. We abandoned these barns and concentrated on the newer ones. We still have to make additional changes. At the time, a lot of steel was employed for partitions. They had to be painted and they rusted anyway. Today one uses plastic instead, which does not have to be painted and is easy to keep clean, which is important to prevent disease."

The cooperative that had planned to concentrate its dispersed pig-raising operation had indeed done so (albeit later than originally planned), had mechanized it, and had fired two of its four workers. Another cooperative has moved in another direction. Because its dairy operation is still losing money even after increasing the productivity per cow per year from a little under 6,000 to 7,650 liters (well over the 7,400 liters the EU figures are necessary for dairy farming to be profitable), enabling the cooperative to reach its quota with a smaller number of cows and fewer workers, and making such innovations as free stalling, the present manager, who came from livestock production, has decided to invest an extra million deutsche marks in a new milking facility that would reduce labor needs in the dairy operation from ten to eight workers. At the same time, however, the cooperative decided to temporarily stop raising pigs until the price for pork recovers. The manager's moves are indicative of the dilemma farming cooperatives face in eastern Germany. On the one hand, she is keenly aware of the pain of unemployment and tries her very best to maintain labor-intensive operations such as dairy farming, even though her predecessor wondered out loud whether it was a good idea to continue raising milch cows. On the other hand, she battled with the BVVG, from which the cooperative rents much of its lands, to allow her to stop raising pigs. The cooperative had promised to raise pigs as part of the original design in order to assure the maintenance of a certain level of employment. The manager and her predecessor argued that the BVVG was holding on to antiquated notions of promoting employment by dictating land use schemes that the European Union had long abandoned in its efforts to progressively accommodate free market forces. She had finally been able to convince the land trust authorities to renew the land lease contract by promising to reconsider the cooperative's decision when market conditions improved.

As in crop production, then, the dual demands of profit and employment had to be weighed in the context of the demands of the local community and the larger EU and world community. In the GDR, cattle raising was geared toward the state-channeled export of animal products to earn scarce hard currencies, even at the cost of ignoring optimal ratios of animals to available inputs. The new economic order brought the force of supply and demand directly to bear on certain barnyard species such as sheep and pigs and, in the case of milch cows, replaced GDR dictates to maintain specific minimum levels of production with production ceilings. For the unregulated production, personal arrangements with local butchers for animals grown to specification could help offset competitive pressures, while milk was sold to oligopolistic regional processors.

Markets

The problem of losing markets described in the introduction to this section continued to plague farmers in the east in the mid-1990s. The processing plants that had been established were often spaced at a considerable distance from one another, increasing transportation costs and creating oligopolistic situations. Thus, a gigantic sugar factory in Sachsen-Anhalt opened in 1993 was the sole outlet for sugar beets for a large area. When technical difficulties led to a two-week delay in the plant's opening, a large proportion of that year's bumper crop lay rotting in the fields. We shall discuss it in more detail in chapter 8. For other crops, like wheat, channels of commercialization already functioned efficiently. In fact, cooperatives had been able to command high prices because of the superb quality of wheat in this dry area and the large quantities buyers could acquire from a single source, which caused western German farmers to fear that *they* would be marginalized. East German farmers found it difficult to penetrate and become accepted in western commercial wholesaling chains that served the local stores. Local goods, even of a higher quality, could not be marketed. In addition, the farmers were still often bitterly disappointed by the low prices offered by west German buyers. "They offered us 0.50 to 0.80 DM," one cooperative manager complained, "for a kilo of pork, which was then resold in the west for 4 to 5 DM. Since there was lots of pork and it was fat, the important thing was to get rid of it. So the Old *Länder* profited mightily. That boom has now come to an end too."

In common with farmers in other industrialized nations, they also felt their worth diminished both in terms of the amount paid relative to the cost of production and commercialization, and in terms of their general social standing in the society as a whole.

By 1999, marketing outlets had stabilized. The cooperative managers continued to enjoy stable arrangements for selling grain, and no one was complaining about the monopolistic sugar factory any longer. After all, while prices of most other commodities had dropped, the sugar price, at least for beets within the quota, had remained stable. Also, as one informant explained, playing off one client against the other brought limited returns, as he had found out when he switched from one dairy processor to another. Some of our informants were improving their processing and storage capacity in order to enhance the quality of their produce and to become less dependent on market fluctuations. Thus one cooperative increased its capacity for drying herbs such as marjoram. It was also adding a fourth large storage hall for onions, one of its specialities. It also planned to install labor-saving technology in the facility. As the manager explained, "It will make it possible to store the onions rapidly after they are harvested. That often still takes too long. Quality suffers. Quality is crucial to be able to place the onions on the market. One can't sell bad-quality ware." Some of the onions have been sold to places as far away as England, Denmark, and Estonia. However, the largest market by far is in Germany. The cooperative, like 80% of the onion producers in eastern Germany, has joined a marketing cooperative, with headquarters in Thuringia, which markets the onions. Although the growers have appointed a Dane, who formerly dealt in onion seed, to run the organization, it does not appear to be dependent on western interests.

While the same cooperative has made stable arrangements with two factories, one in eastern Germany and one in western Germany, to sell its herbs, the cooperative that sells herb oils has had greater difficulty maintaining old and establishing new outlets. "We have had to decrease production to 35 to 38% of what it had been in 1990, when we sold 4 million deutsche marks' worth. The drop since 1993 has been 25%. In 1991, there still were factories that had survived from GDR times, but they went down the drain. No one wants to believe it. I don't blame them for it. When someone wins a war against someone else, the victor exercises his power. They took over the fish-processing industry and after one year they said, 'That's it.' They were the ones who controlled the market. That's what the market economy is all about. We are still struggling. It goes up and down. We are still working with some of the firms, but these firms are always relieved when they have survived another year. There is a registry here in Germany. To be included in that registry costs a lot of money. But even if you are inscribed, that doesn't mean that you automatically can deliver. You are called up and then the question is who calls you up. The system makes it impossible for the small producer to reach the market. He can't keep up. The large producers or dealers will order, say, 100 tons. These lists are the large-scale distributors such as Edeka. In the GDR, the distributors acted independently from the retailers. The retailers bought from the wholesalers who made arrangements with the producers. There were no lists. Here it has to be direct, and a small or medium-sized industry like ours can't compete. We would have to be involved in the entire logistics. In Germany there are no storage facilities. The whole production is on the road, in Mercedes trucks, for example. That is the problem. Storage facilities are expensive but better environmentally."

Much of the remaining outlets for the firm's production are still dependent on links established during the GDR who continue their contracts because they do not want to change well-established and preferred recipes. The soils and climate of the area produce a marjoram with a distinctive taste that cannot be matched by Egyptian marjoram. But, according to our informant, the old-timers are dying out and their successors often substitute cheaper ingredients. "We used to deliver dill to a factory in the Spreewald [near Berlin]. But then a new manager came along with new ideas and he buys from Bavaria." Thus, although he does meet new clients at food fairs who buy 20 liters or so and has an arrangement with a firm in Denmark that makes packaged soups, marketing is an uphill struggle. Access to stable markets for conservative products and major difficulties in accessing extraterritorial markets for specialty products, then, explain, in part, the decrease in innovation in agricultural production.

Reducing the Workforce

An editorial in the *Neue Landwirtschaft* (Paasch 1994:7–9) calculates that between 1990 and 1993 an average of 25,000 workers were employed by cooperatives that were not actually needed. Although these figures hide the fact that, in 1990, the LPGs would not have been able to rationalize their operations by getting rid of surplus workers, since they were required to give their workers several months' notice, the fact remains that cooperatives have provided far more jobs both in absolute numbers and in relation to the surface cultivated than other types of farms.

The reduction of the workforce has indeed been painful. According to a study by the Forschungsgesellschaft Agrarökonomie Berlin, the workforce in agriculture was reduced from 14.5 workers per 100 hectares in 1989 to 3.6 in June 1992, in the new *Länder* (from 14.4 to 4.0 in Sachsen-Anhalt), resulting in a reduction of 87.6% from the 1989 levels (Schmidt 1992:14). Such trends resulted in an overall decrease of employment in agriculture from some 850,000 in 1989 to 157,000 (Laschewski 1998:57) in 1995, a reduction of over 80% and a further 28% reduction by 1999 to 112,600 (Bundesministerium für Landwirtschaft). To give an example, one cooperative started out with 115 workers after the initial subdivision of the LPG. By the end of 1991, that number was further reduced by 20 when its orchards and the labor-intensive potato production were eliminated. At the beginning of 1994, only 39 were left and the feasibility plan (*Konzept*) called for a further reduction to 25 persons (including managerial staff) by 1997.

Women were particularly vulnerable. Thus, in one cooperative, only ten or twelve out of a hundred women remained employed. Another cooperative also reduced female employment from about 50% to between 15% and 20% of the total workforce. If they had not continued with the labor-intensive herb production, the ratio of women to men would have been even worse. The same informant estimated that the average female share of the workforce employed in all the smaller units into which the original collective was subdivided was reduced to only 10% of the total.[20] The severity of these reductions in female employment was mitigated only by the preference for seasonal employment conceded to members' wives in some cooperatives.

In addition to the fact that cooperatives now hired outsiders to provide many of the services for which they had their own permanent workforce, many of their former activities had ceased altogether. Although some cooperatives still served an elaborate breakfast at least to the staff members, lunch was no longer served, obviating the need for a kitchen staff. Day care facilities were closed and the liaison persons with the community were removed. The reduction of the herds both because of decreased demand (especially for pork) and enhanced productivity (milk cows) led to a commensurate decrease in the workforce. The ratio of tractor drivers to the surface of land decreased as a result. Thus, while one LPG used to employ 30 to 40 tractor drivers for the 3,000 hectares (a ratio of 10 to 13.3 per 1,000 hectares) corresponding to the part now held by one of the successor cooperatives, the latter cultivated 1,800 hectares in 1994, with 12 to 13 drivers (a ratio of 6.66 to 7.22 drivers per 1,000 hectares).[21]

The fact that, unlike their west German counterparts, many of the workers had skills that were applicable in other economic sectors facilitated the transition. They ranged from economists to technicians, computer experts, and construction workers. They also tended to be more educated. Thus, 2.9% (versus 0.2% in western Germany) had university degrees, 6.8% (versus 3.8%) had trade school diplomas, 6.8% (versus 4.4%) had a degree as master craftsmen or farmers, while 74.2% (versus 82.9%) were classified as skilled, and only 9.4% (versus 12.1%) were classified as unskilled workers (Hartmann 1995:78–79).

In spite of the concerted efforts of former LPG managers to place their workers, rural unemployment became the single most important negative result of the *Wende*. Schmidt (1992) found that 22% of the workers displaced from agriculture by 1992 retired or took advantage of government programs for early retirement (i.e., at age 55), 15% to 16% found employment outside agriculture (including associated services), 5% to 5.4% commuted to the old *Bundesländer*, 14% were employed in workfare programs set up by the German government, and 16.5% remained unemployed. These figures roughly corresponded to overall unemployment in the new *Länder*, but they hide the fact that many of the workers employed in agriculture were not employed on a full-time basis. Full-time employment may have been reduced from 848,000 to as little as 190,000 (Schmidt 1992:14).

As the above figures show, attrition through retirement and early retirement without replacement was the major means of reducing the workforce. We are not aware of any farm workers and of only one manager who did not take advantage of the early retirement program. When asked what they considered the most difficult aspect of the transformation after the *Wende*, one cooperative manager answered, "The fact that many don't have any work any longer. That is the worst thing that can happen. Even people who took early retirement would have liked to continue working. But we told them that they had a unique opportunity; for those who reach the age of 55 this year no longer are eligible for the program. 'Consider it carefully. We don't want to pressure you, but the risks are far too high. If something should happen, you will be on the street.' So all said, 'Okay, you are right.' Now they are slowly becoming used to life without work, but they are not particularly happy about it." Another cooperative manager, who was over 55 himself, also felt deeply for his older colleagues. "You must understand, I took over the LPG in 1964 and have worked with these people since then. How could I tell a colleague who is 57, or 52, or 54 [that I won't hire him]? After all that time, I know what skills he has, so why shouldn't I tell him that we are able to give him a job? . . . I have experienced old persons who find themselves taken out of their professional life who have become, not physically, but psychically ill. They begin to hide. They believe that they are useless. One has to use extraordinarily gentle persuasion (*mit Engels Zunge)* to move them to assume any sort of responsibility at all. A lot of people have been destroyed in this manner. Those who are used to the existence of unemployment can get used to the thought [that they might be next]. But our people have lived without this fear for forty years, and now suddenly. . . ." In addition to the concern the managers feel for maintaining employment, this example shows the deep identification of eastern Germans with their work. The loss of jobs after the *Wende* signified more than a reduction of income; it meant an abridgement to their rights to self-fulfillment, a loss so traumatic that even an offer of temporary employment could not restore their self-confidence.

Retirement and special early retirement programs still did not reduce the workforce sufficiently. Especially in the beginning, many workers left the LPGs on their own for jobs in construction or in the west. One cooperative manager whose workforce was reduced from 360 to 80 (with an additional 55 in an associated construction firm and 1.5 in a daughter firm engaged in commerce) said, "We were only forced to dismiss two or three persons because they were drunkards or lazy, all the rest left on their own. The people were blinded. They thought heaven had come to earth and they had to hurry there. Many of them are now unemployed again. But I let them go and then said, 'That's the end.' Fortunately, all the lazy ones left, those who created difficulties for us."

Another cooperative manager concurred. "Many said, 'We can earn more and more easily in the west.' That wasn't just a handful. At first, some of them earned twice as much. But of those who left at the time, 80% are now unemployed. They really took advantage of them there. They earned 1,000 DM before taxes in 1991 here and 3,000 DM there. That was a fantastic sum for us, and yet it was just half of the amount westerners were paid. When they realized this and demanded more equity, they were dismissed."

Individuals whose families had brought land into the LPG were more likely to be kept, while those who had engaged in now obsolete activities such as sheep raising were let go rather than retrained. Since unemployment was rampant, cooperatives were often able to retain their most hardworking and most versatile workers, even though some deliberately kept wages very low right after the *Wende* to induce workers to find jobs elsewhere. On the other hand, the strategy of getting rid of as many workers as possible, regardless of their potential contribution to the farm, could also backfire. Some of the most responsible workers had been those in the age bracket that was induced to retire early and individuals with other options might also have been more productive than those without.

For a while, a number of workers could be employed in special work programs. Communities and former LPGs could submit proposals to the employment office and such activities as clearing rubble, planting trees, and (with the assistance of the cooperatives' regular crews) even repairing crumbling buildings, could be undertaken under the supervision of one of the workers who was accountable to the authorities. In one cooperative, up to forty persons out of a total initial workforce of 192 were engaged in such activities.

As mentioned earlier, in 1994, cooperatives also engaged their regular workers in all sorts of demolition and construction activities. Some of these activities were geared specifically to reduce unemployment. As one cooperative manager put it, "[Many rural workers] are unemployed and receive money from the state. That is not the case here. We do not want it either. We believe in looking for work for our colleagues. [We are spending money on reconstruction] because it is the right thing to do, instead of saying, 'Well, we in management have work after all. We are sitting in the warmth, but you should go home and be without work.' It's not like that here."

As we have seen earlier, new technology obviated the need for large workshops. Our informants were vacillating about whether to undertake their own repairs at all. Most, however, continued to employ mechanics. In 1994, one informant said, "We didn't fire our craftsmen. Some became independent or left the farm because they could earn more elsewhere. I would say that if one takes into account the subdivision of the original LPG, there were about double the number of craftsmen. But most of the mechanics stayed on, because the situation was not as good as in other trades such as construction."

In 1999, the question about whether to continue maintaining a workshop and a stock of at least the most essential spare parts became even more pressing, because by then much of the older equipment had been replaced. One cooperative manager commented that one could never be assured that spare parts would arrive in a timely manner, which was crucial during the harvest when time was particularly precious. "For that reason it is also good if one has enough mechanics and whatever else is necessary, so that one can get out of a tight spot." She conceded, however, that "It is probably not as effective (to do most of the repairs yourself) than if you have them done by an outside firm. But there is always the hidden aim to provide employment for people as long as something comes out of it." According to this informant's predecessor, continuing to maintain a brigade of eight mechanics has the additional advantage that they can be used flexibly wherever they may be needed. They provide "a buffer in periods of peak activity, when there is a flu epidemic, or when there are problems in the dairy operation." Similarly, another cooperative continues to engage three mechanics. During the summer season they all double as tractor drivers, while during the winter, one of them repairs whatever is broken in the barns, while the other two go into unemployment until they are needed again in spring. However, since one of the latter is an electrician, he jumps in whenever he is needed. All overtime incurred during the growing season is repaid in the form of paid leave time in December. In contrast, a third cooperative, the one that has stopped raising livestock, has decided to reduce the number of mechanics it engages from two to one and has most of the repair work done by outsiders. Two additional mechanics lost their jobs when the dairy operation was shut down. Finally, a fourth informant mentioned repairs as one of the few realms in the farm where additional savings could be made by having outsiders do the work. It is indeed doubtful whether the cooperative that still engages eight mechanics will be able to do so indefinitely. As the present manager put it, "The scissors (between cost and profit) are opening wider and wider and so we should really fire another five to eight workers in the next two years."

Now as in the mid-1990s, whenever possible, the workforce is reduced through attrition. While the original early retirement program has ended, a new one, geared particularly toward farmers in western Germany, became operative. A 55-year-old farmer who had worked on the same farm for three years and in agriculture for ten and was willing to close his own farm could retire with full benefits. In the east, cooperatives could leave land fallow instead of closing the farm. Thus one cooperative retired fifteen workers and fallowed 80 hectares of land. However, since there was no stipulation of how long the land had to remain fallow, the land was planted again the following year.

Management also substantially reduced its personnel, although the reduction was uneven. One cooperative that engaged thirty persons in administration only had nine in 1994, albeit with only half the surface to cultivate. Another lowered the members of the executive committee from twenty to five. "[The large number before the *Wende*] had to do with democracy," explained the cooperative manager. "There had to be women, young persons, and party representatives in the executive committee. Five are certainly still too many. The cooperative law prescribes two, which would also be enough."[22]

Office staff was also reduced. For example, one cooperative reduced the number of bookkeepers from twelve to four and a half thanks to faster computers and the fact that they no longer need to provide figures to the central state agency that formerly did the final accounting. Another cooperative manager painted a similar picture of a change from minute accounting practices to less attention to detail. "I worked as an economist in agriculture for many years. The farms had a larger administrative staff, but they could provide information on everything, including cost accounting. At the end of the year we could say exactly how much it had cost us to produce 100 kilograms of wheat and the expense of producing sugar beets or milk. Now we only have a few individuals in charge of bookkeeping. Small operations often do not do their own bookkeeping at all anymore, but have their tax consultant do it for them. We now have an accountant and his assistant and the secretary." As in the past,[23] cooperative administrators, unlike private farmers, tended *not* to be engaged directly in field work, except in a pinch on weekends. In one cooperative, there was no direct involvement. The manager told us, "It would be possible, but we don't do it. One [of the persons on the executive board] is a mechanic; he is a master craftsman and is in charge of our workshop. Another is in charge of sales. Also, there are enough problems every day that require attention. It would be pointless for me to sit on a tractor while other things remain undone. It is important to maintain our irrigation system, buy plots of land, plan investments . . . someone has to attend to these matters constantly." In another cooperative, there was only token involvement. The cooperative manager explained: "We don't have to [drive tractors]. We have enough workers, but we do drive the harvester from time to time on weekends, more for fun and out of a sense of connectedness. Also, during the harvest, my colleague is out in the fields from dawn to late at night so that everything works properly." However, in some cooperatives, some members of the executive council were also more actively engaged in agricultural tasks. Thus, in one cooperative, the head of livestock production, a woman, also worked in the barns and the field supervisor also drove tractors, cranes, or whatever was required. Only the accountant and the general manager were not directly engaged in such tasks.

During our second visit to Bernburg, in 1999, we noted some evidence of further reduction in managerial staffs. Thus, the general manager who replaced one of our informants, who had retired, also continued to head the livestock-raising operation. During our interview, she and her predecessor argued about whether it was more or less work for her to play these two roles now than it had been for him to act only as general manager during the transition. While she argued that the executive committee members now had more obligations, he pointed to the reduction of the number of barns from 21 to 3 and to a decline of the workforce from 120 to 40. She countered that it didn't make that much of a difference whether one headed a workforce of 120 or of 40, to which he replied that they had no computers and were engaging in selling farm equipment as a sideline in the beginning. He reminisced that everything was complicated because it was new. And then there was the problem of who owned what land. All this was work that no longer had to be done. She conceded the point that there were indeed fewer major problems, but the decisions that remained were all the more delicate. When one of us asked whether these delicate issues included the fact that workers had few alternatives, she said, "Yes, that is part of it. When one fires someone, one can be 100% assured that he won't get a job again. Most of the industry has been wiped out and we were, after all, a highly industrialized state. Today, if young persons want to work they have to leave for the old *Bundesländer*. For, while we have unemployment ratios of 22% to 23%, they have 17% to 18%. That makes a big difference."

Wage Levels and Working Conditions

The level of agricultural wages, which was roughly commensurate with those in other sectors of employment in GDR times, was lower than in such sectors as construction in 1994. In the cooperatives studied, typical average hourly wages ranged from 11 to 13 deutsche marks, or a monthly wage of 1,300 to 1,500. In summer, a tractor driver could earn over 3,000 deutsche marks a month, working ten to twelve hours a day for twenty-six to twenty-eight days rather than the usual twenty-two. Barn workers could earn similar amounts. Government-mandated benefits increased average compensation to some 20 deutsche marks. In addition, workers had the right to up to six weeks of sick pay and twenty-six days of vacation (for workers over 40 years of age) per year, a figure that was considered low by western German standards. One cooperative manager figured an average of 35,000 marks per year including benefits. In contrast, the seasonally employed women earned only 8 deutsche marks plus 2% to 3% for insurance. According to one informant, managers earned about double of what a top tractor driver earned, a ratio similar to that in GDR times. Cooperative managers resisted the temptation to hire foreigners as seasonal workers, even though Polish or Czech farm hands could be hired for as little as 5 deutsche marks per hour plus 2% to 3% for insurance.

A cooperative manager calculated that a family could not live on only one farm worker's income. There would have to be two incomes or one income and one person receiving unemployment benefits. He figured that a family of four would need between 2,200 and 2,500 deutsche marks per month to live more or less well in eastern Germany compared to the 3,500 to 3,800 deutsche marks his western German relatives figured would be the minimum needed where they live. However, most of the cooperative members had their own houses and therefore did not have to spend the typical 500 deutsche marks for rent (for a new 68-square-meter apartment).

By 1999, cooperatives had increased wages somewhat. Thus, one cooperative paid an average of 14.50 to 15.00 deutsche marks per hour, slightly below the 16.00 advocated by the union; another paid between 15.00 and 16.00.

Working conditions themselves were changing only slowly. Whereas before, sugar beets were often still hoed by hand, hand labor in the fields had been eliminated entirely by increasing the use of herbicides. But for some workers, the demands may have even increased. For example, mechanics, who were often largely exempt from engaging in other agricultural activities in the past, now also had to drive tractors or other farm vehicles. For other workers, there were improvements to the extent that old technology had been replaced. As one cooperative manager explained, "The profession of tractor driver leads to a greater deterioration of the body. It is very strenuous. The body is constantly subjected to shaking, much more than with truck driving. The noise level is high and, in summer, there is the heat and the dust. Last year, we bought a new harvester where the driver sits like in an automobile. It requires little strength to operate; everything moves easily; one barely hears the machine; it comes equipped with a radio; and the dust doesn't penetrate the cabin, which is air-conditioned. Our machines had none of these nice features. . . . [So working conditions have improved] to the extent that we were able to buy new equipment. We have one new harvester and two new John Deere [tractors]; but other workplaces have not changed."

The Cooperatives' Relationship to Communities

The disputes surrounding the debts incurred by LPGs before the *Wende* and the unwillingness or inability of the various actors involved to take on community projects, has altered the relationship between cooperatives and the communities in which they are located. The cooperatives have attempted to request compensation for the investments in roads, but so far without any success. As one cooperative manager commented, "At present, we are thinking about how we could get at the wealth that was demonstrably created by us in specific villages and cities, but which others are claiming as their own. But whether we will succeed or not is an entirely different matter. For example, we submitted a blanket request to Mr. Weigel (the finance minister). But he replied with a friendly letter, stating that there was nothing to be had. He wrote that there are laws that allow property to be claimed under certain circumstances, but not in our case. We can prove certain things and we will have to see whether it makes sense to take any further steps or not. Do we have the strength to take up litigation? If we don't, we will never receive anything. Just to give an example of an investment, we used to build roads. The *Treuhand* has given title to these roads to the town administration. But the asphalt surface is ours. It's worth more than a million. We politely notified the town managers. But they have not replied yet. It has been many years now (since we built and maintained the roads), but these rural roads are in better shape than the communal ones. So we feel that a certain sum should be our due in one form or other. At the very least, the communities should take over their maintenance."

In the meantime, the roads slowly disintegrated. As the same informant described, "Last year we still got a road scraper for 5,000 or 6,000 deutsche marks to fix the gravel roads, so that they would at least shed the rain and last for a while. We did that for those who share the road as well. But we can no longer afford to do that." The resentment against the authorities was echoed by another cooperative manager. "After the

Wende, people were upset that there were signs on the roads we built and maintained, ONLY FOR AGRICULTURAL VEHICLES, even though there are even more such signs in the west. So they were taken away, and construction vehicles, heavy trucks, everybody used the roads. Now none of them are usable any longer. They are full of enormous holes. I said, 'Our tractors have large wheels. We are not interested. Finished, we won't spend another penny.'"

In spite of these disputes, the relationship between the cooperatives and the surrounding communities continued to be quite amicable, for the cooperatives provided more work than the individual farmers. As one cooperative manager put it, "The personal ties remain even though the assistance (we gave) to the community is gone." "It would be impossible," added the co-manager of the same cooperative, "if we transported things for the community. That would be considered a trade for which we do not have a license. Someone in the community who is licensed for that trade could complain to the internal revenue service. We can't even do such things as readily as we did before for our own members. Today, everybody keeps a close watch over mistakes committed by others." The major effect of the *Wende*, then, was a disengagement of enterprise and community, a fractioning that characterized many other aspects of life in eastern Germany as well. Many commented sadly on this loss of common goals.

Prospects

The statistics on agricultural cooperatives show a progressive stabilization of their numbers since the conversion of collectives was completed at the end of 1991. Their numbers continued to decrease between 1992 and 1998. That decrease slowed from 10.2% between 1992 and 1995 to 7.4% between 1995 and 1998 (Laschewski 1998:57, 193; Agrarbericht 1999). A similar stabilization could be observed in the average amount of land they cultivated: while cooperatives continued to lose land to individual enterprises, this loss decreased over the years. Given the prevalence of plant closings in other economic sectors, "the stability of the successor enterprises to the LPG is quite spectacular" (Laschewski 1998:146). To be sure, this stability was achieved, in part, by compromising the managers' stated ideal of maintaining as many jobs as possible. As we have seen, between our two visits, there have been further reductions of the workforce. In the extreme case, this has meant the abandonment of entire branches, such as dairy farming and—at least on a temporary basis—raising pigs.

Indeed, one of our informants, a cooperative manager who has retired, wondered whether the ideal was upheld by management on its own or whether it was merely responding to pressure that was brought to bear by the members. His successor denied that the members engaged in livestock production would have the votes to act as an effective pressure group, to which her predecessor replied that she had not taken into account the fact that there would be changes in crop production as well if the dairy operation were shut down. Thus straw would not have to be harvested, and fodder crops would not have to be planted. Both agreed that crop production was subsidizing livestock production and that the latter had been in the red all along. The pain of having had to fire so many co-workers also comes through in the following quote of another cooperative manager: "I can't hide the fact that whether it suited me or suited others or not, [putting the cooperative on sound financial footing] did not happen without decreasing personnel. I was also a proponent of maintaining as many jobs as possible, but it isn't easy. We had worked with these people for many years. We live in the same village here or in the neighboring one. One can't just play a dirty game with them. It's not like in any old large firm where management is located in an entirely different place and doesn't have any contact with the workers. That is not the case here. One has to harmonize everything and that is not easy. It requires a type of management that is not just familiar with figures and enters into speculative deals but has to know how to persuade people."

In conclusion, the cooperatives still hold on to the socialist ideals whereby priority is given to livelihood considerations for all over abstract principles of profit or economic aggrandizement of the managers.[24] They have done this at the cost of maintaining a modest image and relatively low compensation of their managers. In fact, although their remuneration measured by the cost of living may have risen since the *Wende*, the managers no longer have access to perks such as access to vacation homes and, perhaps, easier access to building materials for homes and scarce goods.[25]

CHAPTER SEVEN

The Eastern Independent Farmers

Asked why he had decided to establish a farm of his own rather than work as an employee for a western German farmer, an individual eastern German farmer explained, "It isn't a secret. . . . Just after the *Wende* someone came here from West Germany. He came quite often in fact. We didn't really know what he wanted. He had an enterprise over there and he had 250 hectares here. He kind of thought that he would work on his enterprise there and he would hire someone here. That would have been us. We didn't want that! So we did it on our own. We are three men and we can manage."

Explaining the impediments placed in his way by the collective to which he had originally belonged when he decided to set up his own farm, another eastern German said, "We had some ten owners [who decided to rent the land to us], each with 20 hectares, making a total of 200 hectares. But the pieces were not contiguous. So the cooperative told us, 'It is difficult to make land exchanges. Let's do it differently.' They gave us 200 hectares [that they had been renting] because, after all, we had talked to the owners. They said, 'These and these fields amount to 200 hectares. But you will have to make arrangements with all the actual owners of the [component] parcels. Well, included in this land is a field of less than six hectares that belongs to 121 different owners!'"

This chapter is about the former members of agricultural collectives who decided to leave the collective and establish farms of their own, rejecting job offers of westerners who took advantage of relatively low lease rates to supplement their generally small farms back home and bucking the resistance of cooperative managers who placed all sorts of obstacles in the path of the wayward members to dissuade them from jumping ship, and reducing the cooperative's holdings by competing for the same land.

The option of establishing an independent farm was the one favored by the authorities, but, as we have seen earlier, relatively few former members of cooperatives took this path. Part of the reason for this fact is certainly the legacy of the agrarian reform of 1945. That reform, by giving the death blow to the former latifundia (large estates), paradoxically also provided a safeguard for cooperative agriculture. For it provided too little land for the beneficiaries and their descendants to start up their own farms without gaining land from other sources, making the option of joining a collective a viable, if not the only, choice. This fact made it difficult to establish private farms after the *Wende* for the beneficiaries of the reform and for the deposed nobility alike.

A major disincentive to establishing an independent farm was the risk involved. Although farmers could count on massive state subsidies that reduced the risk of catastrophic failure, one of our informants, a westerner, was forced to give up farming by 1999. And there were stories of the failure of one or two others who had operated farms in the east while maintaining their residences in the west. As one independent farmer commented, "I talked with many [of my colleagues who decided against becoming independent]. Maybe it had to do with how our system worked before. In the past, others thought for us. We didn't have to think. Everything was automatic. Now I have to make a request here and a request there. If I don't—and in a timely fashion—I receive less or nothing at all. It was different before. Everything was organized in a Prussian manner and that was that. Many are satisfied if they get a fixed income each month. Why should they incur risks? Many were afraid to [give their land as collateral for bank loans]. They said to themselves, 'If I put this down as a security and things don't work out, then the bank may come along and take it away from me.'" The willingness to take on the risk of establishing a private farm was partially dependent on whether one could rely on the contribution of the entire family. As one farmer put it, "For one, I was afraid to take on debt because we had no experience of that before, although, as it turns out, the credit conditions are favorable at present. But, at the time [i.e., when they first started] we didn't see it that way. We were worried about the debt and also whether we would be able to manage. Would the whole family pull together? That's always a question mark. As I said before, we are in the shadow of the Harz Mountains, so I am always dependent on the heaven for rain. So I am thinking of reducing the risk with some animal husbandry.

Nevertheless, the overall risk remains." The normal vagaries of climate and the risks of production seemed less threatening to the farmer than the new burden of debt and the vagaries of financial management which turned out to be less ominous than expected. Indeed, according to Thiele (1998:210–211), because of the high subsidies, a profit of 2% on the total capital invested would have been sufficient to make a farmer's own investments worthwhile. For some investments, profitability could even be negative and it would financially still make sense to invest.

Some of our informants attributed their decision to become independent mainly to the need to secure a livelihood in uncertain times, for the risk of unemployment was and continues to be considerable. As the manager of a cooperative explained, "No one dared to take that step right away. But one year later, a father and his son decided to do so. The father said that he had 25 to 28 hectares of land in the collective and could lease additional land. The son had obtained a degree in Halle and it became apparent that he would not find employment either there or elsewhere, so he said to himself, 'I might as well start a farm of my own.' He is capable and was trained so that he could also operate farm machinery himself. They were the only ones to leave our community."

Similarly, a *Wiedereinrichter* (lit. "reestablisher," as the local farmers whose families joined cooperatives in GDR times and who decided to take up individual farming were called) acknowledged, "I mulled over the decision for a long time. I could possibly have found work as a crane or other equipment operator in construction, [but I could probably not have stayed in my profession]. Education for apprentices in agriculture was cut back, because we now need fewer of them and the training has moved to another location. My son now goes to Kemberg. So I would probably have lost my job. I said to myself, 'Rather than becoming unemployed, I might as well start something of my own.'"

Another *Wiedereinrichter* first thought of taking up agriculture only as a part-time activity to complement his income from his position as mayor of a small village. He was in conflict with the head of the cooperative, whom he suspected of personal enrichment at the expense of the cooperative, and did not want to continue working there. But, since the state abolished mayorships in small villages, he took up agriculture as his principal occupation. Yet another farmer intended to expand his small private farming operation to truck driving or acting as a farm manager for a western German who was planning to operate a farm without permanently moving from the west and who offered to hire him and two other local farmers.

However, the fear of unemployment was never the only reason for becoming independent. Indeed, one of our informants could have followed the example of the great majority of his age mates and gone into early retirement. The local farmers who decided to leave their cooperatives and set up their own operations tended to be from families who once owned larger farms, owned their own houses and sometimes farm buildings, and (in the case of older farmers) had assured successors trained in agriculture or were in the process of obtaining such training themselves (the latter constituting a formal requirement for state subsidies). They had joined collectives late and had maintained larger than usual private animal husbandry operations during communist times. These local farmers often established local marketing outlets.

The adaptations of the eastern *Wiedereinrichter* to the conditions after the *Wende* were both similar and different. All of our informants had a background in agriculture. As we have already indicated, all six of our eastern German informants with individual farms came from farm families and were thus true *Wiedereinrichter*. All of our informants had already raised pigs before the *Wende* and probably all raised hens. Four—one of whom even produced milk—also fattened cattle. They were thus all familiar with the basics of raising animals.[1] In addition, four had held managerial positions in the collectives to which they belonged and had received specialized training in one or more aspects of agriculture. In one family, a son and daughter-in-law were also highly educated. Two other *Wiedereinrichter* and the son of one these two had even taught agriculture. Since a minimum of one year of full-time schooling, or three winter seasons, are required to operate a farm and train apprentices, both father and son in one farm, and the son in the other farm, where the owners were semiskilled farm hands, were catching up on their formal training at a nearby agricultural and technical school. Present-day training is more general than it was in the GDR, where crop and livestock production was separated and where large cooperatives hired more specialists. Nevertheless, none of our informants complained that their one-sided training before the *Wende* had placed them at a disadvantage. Instead, they stressed how they used their specialized skills in their new operations.

A special case of privatization was the transformation of one of the rare remaining Type I cooperatives, that is, cooperatives where members only pooled their land and not their animals. In the sole Type I cooperative that had survived the final collectivization drive in the 1970s in the Bernburg area, the small cooperative with only seven members described in chapter 4, four of the members decided to retire, while the remaining three formed a common-law partnership (*Gemeinschaft bürgerlichen Rechts)* in which each member ran independent operations but pooled some of the machinery.

Only one of our *Wiedereinrichter* informants had not worked in an agricultural collective before the *Wende*. This farmer, who had always raised bullocks, pigs, and horses and had cultivated plots of land that were too small to be of interest to the cooperatives, owned land on the outskirts of Bernburg which he was able to sell to a developer to finance the establishment of his own farm.

To some extent, the thought of being subordinate to westerners also irked the would-be new farmers, as was the case in one of the examples introducing this chapter. Others gave similar reasons for striking out on their own. "I wanted to show them that we could do it too," one recounted. Another *Wiedereinrichter*, Mr. H., decided to become an independent farmer for similar reasons. "After the *Wende* I worked in the cooperative and was in charge of plant production. I was the only person who owned land in the cooperative (actually only 5 hectares); all the others were just wage earners. That makes a difference. If you want to engage in a joint endeavor, you have to be equal. A second reason was that I didn't like the fact that people are coming over from the old *Bundesländer* to become farmers here while we just sat back and watched. I wanted to show them that we also know how to do things."

In the period between 1994 and 1999, very few new independent farmers were able to establish themselves in eastern Germany. In the area studied we heard about only two or three such cases, only one of which involved an easterner. The latter, whom we have already encountered in chapter 5, applied to the *Land* to take over a 300-hectare demesne in 1994. The arrangement with the *Land* includes taking over a beautiful old house and sharing its upkeep with the *Land*. Although the lease is for eighteen years, the idea is that the lease could continue within the family. Indeed, one of the daughters, who is studying agriculture, is a likely successor. His wife also became involved in the farm after losing her job running the store owned by the cooperative. She does the accounting. He took charge of the demesne with the consent of the cooperative of which the demesne had been a part. The cooperative had also applied for permanent management of the demesne, but was told that only individuals could obtain it. At the time, the farmer had intended to remain as manager of the cooperative that had cultivated the demesne land and have it continue to till the land for him.

Although he is still a member, in 1995, he had to make the decision to relinquish his position, sooner than he had planned, as a condition for obtaining the demesne. He continues to maintain a close relationship to the cooperative. While he does the planting and tending of the crops himself, he has the cooperative do the harvesting of most of the crops and the sugar beet harvest by someone else. He figures that that is more

FIGURE 7.1. House for the renter of an old demesne.

economical than purchasing a combine and other equipment himself. He has no intention of making any changes in this arrangement. However, the close relationship with the cooperative he used to manage has come at a price. He decided to abstain from applying to take over a second demesne that lay within the cooperative's domain so that the cooperative could retain it by establishing a GmbH to manage it.

Individual farming enterprises were frequently defined as family enterprises or were expected to become so in the future after a period of consolidation. The example of the Feldmanns we discussed in chapter 2 shows the importance of a family history in agriculture in the decision to become independent. Taking the risk depended on being able to rely on family members for advice, financial support, and to supply a workforce. Although the older generation could be skeptical about plunging into what seemed like a very risky venture, they did stand by with advice, and invested the land they had formerly owned as well as their animals, cash, and labor power. Wives and other female members of the family (mothers, daughters, and daughters-in-law) continued to work off the farm until they lost their jobs, to provide the cash necessary to keep the household afloat. An example of this is a newly independent farmer who said, "During the really tough times we have lived on the money that my wife earned. The money that I earn I try to reinvest in the farm in order to pay off high-interest-rate loans, so that the enterprise becomes more effective. When the farmer has money, he buys a tractor. When the question arises whether to buy a car or a tractor, I always opt for the tractor. However, we are still modest in our wants." At first, the wife continued to hold a job in the city bureaucracy. But then she could only work there for six hours a day and finally, she was asked to leave. For two years, she helped only on the farm and then returned to the city to work part-time in the post office. In another case, a daughter-in-law was provided a state subsidy for the first year of her baby's life so she considered her contribution to the farm as "free." Other members of the family added their pensions, retirement incentives, or early retirement pay and savings to help the farm survive the first critical years.

All the *Wiedereinrichter* were willing to live very frugally in order to be able to reinvest much if not most of the farm's income. The eastern German farmers did not seem to see frugal living as a sacrifice. Rather, they saw it as a continuation of a former lifestyle that should eventually lead to a higher standard of living. That they regarded their present situation as temporary becomes clear from the following quote: "In the newspaper, they said that the eastern farmers earn almost twice as much as the western farmers do. In the first year, I was 50,000 marks in the red. In the second year, I broke even, which means that I was already earning 50,000 marks more. If I have a profit of 50,000 marks in the present year, that means an increase of 100,000 marks [over the first year]. We are much better off now than we were then, but we have not yet reached an acceptable level. If I were to earn 50,000 DM per year, I would be very satisfied. From that I would still have to subtract social security, pension payments, and health insurance." In the meantime, most independent eastern farmers, like their counterparts in other sectors, have resisted western consumerism.

Unlike other enterprises in the East, succession seemed less important in agriculture, although three of the farms were two-generation enterprises and each farmer did hope that the sacrifices of this generation would benefit someone if not in the next generation then in the following one. The willingness to accept risk, the fear of unemployment, experience and training, competition, and familial support and the possible inclusion of family members, especially unemployed females, were among the major factors leading to the reestablishment of independent farms.

Obtaining Land

With the decision to establish an individual farm came the scramble for land. In 1994, the six *Wiedereinrichter* we interviewed possessed between 5 and 28 with an average of 18.5 hectares of their own or family land. But they needed much more to operate a farm. The expert who advised many of our informants considered that a family farm in the area required between 200 and 300 hectares.[2] However, some of our informants believed that their equipment would be more effectively utilized if they had access to some 500 hectares and worked them with three persons.

Since eastern farmers owned only 10% or so of the land they needed and could not possibly aspire to acquire enough land through purchase, they had to seek ways of renting land from individuals, the church, and/or the *Treuhand* to make up the difference. The church became the most reliable source, for it rarely rented land to cooperatives. The *Treuhand* theoretically—and often in practice—gave preference to individual farmers, but, as we have seen, pending or competing claims initially made the agency reluctant to sign long-term leases. Also, our informants complained of interminable delays. One *Wiedereinrichter* described his travails in particularly vivid language. "First I wrote [the *Treuhand*] and, when I did not receive a reply, I phoned them. They told me to appear in person. One was supposed to register in the waiting room and after one to five hours one could talk to someone. No one paid any heed to my written request. It was simply thrown away. Only when someone appeared in person was any interest shown. When I tried to establish the basis for renting the land, I found that they had no documentation for the land in question. Neither did I. When I finally had everything together, I returned. This time, [I was told] that one had to do everything in writing. I submitted my request and simply started to cultivate the land without any permission. In February or March, I finally received a reply that I would receive only 180 of the 240 hectares I had requested. So I returned the rest to the LPG, even though I had already cultivated the land. Then I resubmitted a petition to lease the land. [For, what I had received] was merely a notification that I could cultivate the land [rather than an actual lease agreement]. I only received credit for 100 hectares. It was barely sufficient. I was hurting.

"Thereupon, I drove there again and wanted to pay the rent. Before that, I had tried for a month and a half to obtain an appointment by telephone and had finally succeeded. When I got there, the office had moved to another floor. There I encountered a door with a mail slot. I saw a cleaning woman whom I greeted in a friendly fashion, and she let me in. There were no names on the doors, but I was able to find out which one it was and placed myself in front of it. At 8:10 A.M., the woman in charge came. I was allowed to enter and she showed me all sorts of things. I asked her to write a lease contract for the next two years. She answered that that was impossible because it was not going to be an actual lease. She did give me the price though, a horrendous sum of 4.50 DM [per point] when 3 to 3.50 was an acceptable amount. I really had had it! The telephone rang and no one answered it. People stood in the hallway and no one asked them to come in. I left the door open so that they could enter too. How improper! Shouldn't we have the possibility to have a hearing? If I had the choice, I wouldn't have anything to do with them. They could take the land and do with it what they wanted. In my opinion they do all this so that their jobs will last longer. They don't want to work themselves out of their jobs. The information should be in the computer and so one could just write the contract. Perhaps one would have a preliminary conversation and then everything would be settled. . . . Now I have a contract until September 1994, but according to the person in charge it is likely that I will get a renewal."

There were also accusations that the *Treuhand* had engaged in outright discrimination and dishonorable actions. Persons who had not worked the land were given preference over those who had. A GbR with 400 hectares involved in both crop and livestock farming wanted to take over a former state farm with 250 to 280 hectares and two and one half employees. They wanted to invest 1.2 million DM and provide five long-term jobs, but they were not considered. Instead, a western German with influence in the right places won the bid although his proposal did not fit local conditions and did not even mention long-term employment.

Since they could not obtain sufficient land from the *Treuhand*, the independent farmers had to compete with the cooperatives for land owned by individuals. Cooperatives had the advantage of "historical ties" and a common socialist tradition, for they had been cultivating the land since the cooperatives were formed in the 1950s and 1960s. They could put pressure on their former members to continue leasing their land to the cooperative based on a sense of loyalty. Also, they usually did not allow their members to rent their land to someone else. Often, individual farmers rented from landowners who were no longer active in agriculture or at least not in that region.

Although cooperatives were legally bound to release land formerly cultivated cooperatively when the owners decided to rent it to individual farmers, they did not always cooperate. Their tactics included delay or refusal to engage in parcel swapping and/or in releasing land already under cultivation before the next harvest. Stories abounded of cooperatives that refused to release land at all until legal action was taken. Our own *Wiedereinrichter* informants did not have to take recourse to legal action, but they did have considerable difficulties in their dealings with the cooperatives. Thus one farmer rented 100 hectares of land from the son of a man who was killed in Buchenwald. 180 hectares of his 280-hectare farm were divided up in the land reform of 1949. The remainder became state land. The former owner petitioned to repurchase the land and promised to lease it to our informant. Since the *Treuhand* only gave him permission to lease the land in January, when the land was already under cultivation, the cooperative refused to release it until the harvest was over.[3]

We recorded the travails of another two members of a two-generation farm family, a man in his 50s and his daughter-in-law. "The difficulty was that we were the only ones to start again in the 6,000-hectare area already in 1990," said the father-in-law. One field, the one that served as an example of the tensions between *Wiedereinrichter* and cooperatives at the beginning of this chapter, is, as the daughter-in-law explained, "made up of 500 square meter plots that were given to each home owner in town to make a garden or to tether a sheep." "We had to identify and seek out the owner of each of these plots," continued the father-in-law. "'Oh, the good Mrs. Schmidt has died and her grandchildren live in Hamburg.'" The daughter-in-law continued the conversation. "This is the fourth year that we are working on this problem. . . . And now the state is pressuring us, accusing us of trying to enrich ourselves because we have not paid what in many instances amounts to 10 or 12 DM for rent. . . . All in all, we have 170 lease agreements." "It is impossible to figure the time we have already spent on this," added the father-in-law. "But now it has become easier, because there are more private farmers. Now I can lease 100 hectares from just five owners. Of those 100 hectares, I can cultivate 40 directly and swap the other 60 [to make workable fields]. But because we were the first ones in 1990, they told us that we couldn't make exchanges."[4]

Cooperatives were not the only competitors for land with whom the *Wiedereinrichter* had to contend. Competition with western farmers eager to gain a foothold in the east was equally severe. In fact, while it was often possible to come to an agreement with cooperative managers on how much to offer to landowners for rent, westerners often outbid the easterners. As one *Wiedereinrichter* explained, "There is a man in the neighboring village who came from the old *Bundesländer*. I can't consider him as a friend. I have told him as much. The cooperative and I had the same problems with commercialization and with cash flow. Those who came from the west had money and knew where they could sell their products. So they were able to rent the land away from us." He went on to explain how land rents had gone up because the westerners bid the prices up and how many landowners were cheated out of their money by unscrupulous westerners. He concluded, "Without the demand from the west, we would have had an easier time."

Although in 1994 few individual *Wiedereinrichter* had acquired land through actual purchase, some intended to do so in the future. At the time, actual purchases of more than a few hectares of land would have been financially uninteresting for most farmers, for they would have had to pay between 10,000 and 20,000 deutsche marks per hectare of land, which would have meant some 800 to 600 in annual interest alone, compared with 400 for rent. But one farmer did consider investment in land as a good option. "If the choice is to pay 300 to 450 deutsche marks [per year] in rent or to buy it outright and then it will be mine in twenty years, the latter is better. Otherwise the money is gone."

As a result of all their efforts, the *Wiedereinrichter* we interviewed gained access to between 135 and 397, with an average of 252 hectares of land, which as we have seen is very close to the size recommended by the experts but smaller than what they themselves regard as ideal. All but one of the farms in our sample is categorized as "large" in a study of eastern German farms undertaken by the German government in the

FIGURE 7.2. New home of an eastern German independent farmer.

FIGURE 7.3. Old barn purchased by a *Wiedereinrichter*.

1992–93 season (Top Agrar Spezial, February 19, 1994, 10–13). The average size of farms in that category, which includes farms with farm income over 100,000 DM, was 260.4 hectares.

The size of the farm and the possibility of expansion depended not only on the possibilities of leasing but also on the original location of the house and farm. One farm, located within a historic city, was limited in its activities by city ordinances with respect to noise and the like, but the family was unwilling to move, in part because its central location was crucial in serving and securing clients. In another case selling urban land profitably for development resulted in the move to a nearby rural area where the family could buy and rent sufficient land for a new house and barns.

Expansion meant acquiring unused buildings, part of which may have belonged to a former LPG, or obtaining land on which to build. The frustration associated with such an acquisition was expressed by Mr. B. "I am still trying to invest in a riding stable. I got the money for a building, so I am building a structure to house machinery. . . . I bought part of the land and structure but the other part belongs to the LPG that still uses it a bit. But they are not giving me the permission because the community would have to provide water, sewage, electricity, and so on." The size of holdings, then, was heavily dependent on easement agreements, location, and infrastructure and thus subject to annoying delays.[5]

By 1999, some of the eastern independent farmers we revisited had been able to further enlarge their farms through rental and purchase. Thus the Feldmanns increased the size of their farm from 266 to 342 hectares. They were lucky enough to be able to purchase 35 hectares of land from the BVVG through its program of selling off land at reasonable prices (0.70 instead of the going rate of 1.70 deutsche marks per square meter) to established farmers before that program was frozen. They had to pay the sum in cash. In all, they now own 80 hectares, up from 20 or so in 1994. Like the cooperatives, they purchased much of the land from individuals. One of the plots was purchased from a woman who had just retired and who received only a small pension because she had earned a very low salary. "Her son has moved away and she has to keep up the house by herself," explained the younger Mrs. Feldmann. "She said, 'I can't manage without selling land. I am at the poverty level.' She said that she wanted to keep the small plot on which the house stands but that she wanted to sell the large one in order to be able to redo the roof." "We paid almost double of the going rate for the land," added the younger Mr. Feldmann, "because the owners live in the town and I said to myself that I wanted to be able to look them into their eyes after the purchase." "Yes," continued the elder Mr. Feldmann, "this type of situation has often been taken advantage of and the owners were paid 30 to 50 cents per square meter instead of 1.20." "I depend on the townspeople to buy the potatoes I sell," added Mrs. Feldmann. "Therefore, it was important to keep good relations with everyone." Another family, who had sold 15 hectares of land located near Bernburg in order to build a new house and start their farm, had been able to purchase 16 hectares of land they had previously rented. The total size of their farm had decreased from 296 to 283 hectares, but they cultivated an additional 80 hectares for someone else. The third case, a partnership (GbR) of three unrelated families, was more complicated. After its dissolution for reasons that will become apparent later, our informant sought additional land. Although he had only 90 hectares in the partnership, he was in fact able to acquire 150 more, 17 through purchase from the BVVG.

Machinery

The strategies employed by our *Wiedereinrichter* informants in 1994 with respect to farm machinery varied widely. Although most originally employed old equipment, some now used mostly new machines. Others made do with old equipment purchased from cooperatives. Still others had a mix of both. Some purchased tractors with just enough horse power to do the job. Others preferred to have some reserve power and greater longevity.

Only one farmer pooled equipment with others. The four farmers had been members of a Type I LPG and decided that after having worked together for thirty years, they were going to continue to do so. "Each one of us has his own farm, we then calculate what each person owes, according to the amount of land he has. We have only one large tractor and plow, [namely, mine]. So, when I plow, I plow the land of the others too. In this way I have been able to afford a larger tractor than if I was working only my own land. Similarly, when my colleague harvests, he harvests my sugar beets too. Otherwise each one of us would have had to purchase his own machinery." The same farmer also had a more informal arrangement with another neighbor who cultivated only a small amount of land. He helped him with his tractor and, in return, the neighbor's son drove his tractor whenever he was in a pinch and, since the young man was a first-rate mechanic, he also fixed the machinery. In this arrangement no money changed hands. One of the reasons that pooling is less common in the Bernburg area than in western Germany is the relatively large size of the farms. As a result, such machinery as combines is used to capacity on one's own farm. In

FIGURE 7.4. Some of the machinery owned by a *Wiedereinrichter*.

part for the same reasons, most try to limit contracting outsiders to undertake agricultural tasks to a minimum, and then only for tasks where they would require specialized equipment which they feel they cannot afford yet. Many farmers did belong to "machine rings" encompassing several *Kreise*, which enabled them to make their needs and their willingness to provide services known to a wide membership. Membership dues appeared to be low. One of our informants claimed that he did not even know how the machine ring functioned but belonged to it anyway because, "You never know whether you might need a combine some day, but other than that . . ."

Our informants preferred not to borrow from their neighbors since, as one farming couple explained, "When we borrow something it tends to break down just when we use it. Or it is already broken and we have to fix it first before we can use it. After all, some of the machines can be purchased readily secondhand. We still have a good crew of pensioners who can help us fix things. One can weld, another is a good mechanic. As the saying goes, they can make a Sunday dinner out of an apple and an egg. This is much better than if we had to go to some specialized smith shop. That would increase the costs significantly." In addition, our informants did many of the repairs themselves. Some swore by western technology either purchased new, or, in one instance, obtained from a relative with a farm in western Germany. Others had had good experiences even with old GDR machines. As one of the smaller-scale farmers remarked, "I bought a tractor and a twelve-row harrow for 3,800 D-mark. I also bought a combine for 2,000 D-mark. Since it was in need of repair, I exchanged it for two smaller but serviceable ones. I have worked with them for three years now without having to make repairs."

Finally, the farm manager turned demesne farmer continued the same strategy that he had used when he was running the cooperative. In addition to buying an old but low-mileage Wartburg automobile of GDR manufacture for 2,000 deutsche marks to replace the one that was stolen from the cooperative when he worked there, he purchased a Czech 160 horsepower tractor and a smaller 80 horsepower Russian one that cost him one-third of the cost of a German one. "The Czech tractor is cheap but it does have air-conditioning and a seat with German springs, rather than the original Russian seat. I will certainly keep the Russian one. They last and one can easily repair them oneself. Also, even though things break on occasion, the spare parts are cheap. The other one does the heavy work and will certainly break down sometime. It has 4,000 hours on it and is beginning to go. I hope to use it for 8,000. I don't know whether it will last that long and don't know what I will buy then. But I didn't save on the machines pulled by the tractors. I bought an expensive Swedish planter and an equally expensive plow. I get away with low horsepower for the amount of land. All in all I am in the forefront as far as costs are concerned." The manner in which machinery is bought, used, shared, and repaired then resembles the predilections and the bricolage of former times in spite of the fact that commodities and services are more available in the new world order.

In 1999, our informants all seemed even better equipped with machinery than they had been in 1994. For example, the Feldmanns, who had a 50 horsepower and a 114 horsepower tractor in 1994, added one of 150 that year and another of 130 in 1996, all of John Deere manufacture. But reliance on resources from before is also manifested in the fact that the Feldmanns can depend on a 70-year-old neighbor to repair their machinery, provided no electronic components are involved, and have therefore been able to extend the life of their two older tractors. Even the farm family who was best endowed because they had been able to invest money from the sale of some of their urban land could not resist a bit of GDR whimsy in their technology. In addition to four tractors, including three John Deere behemoths, even the oldest of which looked like new because of the care with which it was treated, not to mention the fact that they unlike other independent farmers in the area were the proud owners of sugar beet harvesting equipment, they purchased a small two-wheel machine to pull a wagon on which they had attached an ancient garden bench as a seat. And the garden gate of their beautiful, large new home had been handmade by one of the sons by welding metal rods and an old carriage lamps together.

As we have mentioned earlier, the seemingly idyllic partnership of the four farmers was dissolved. One of the partners retired and another wanted to force the inclusion of a brother, which would have created an imbalance of power. It only became apparent now how expensive it was to dissolve such an arrangement, for taxes had to be paid on the value added since its inception. Needless to say, the informant who had shared machinery with his partners was forced to purchase the machinery he lacked in order to work his own land. The dissolution of the partnership did not, however, end the reciprocal relations with other neighbors. In the past, unable to use joint machinery for this purpose, he had had to use an old tractor which he could now sell.

Investment

The *Wiedereinrichter* also needed large amounts of capital to establish modern farms. They were reluctant to divulge their annual income. In

FIGURE 7.5. A tradition of making do with whatever is at hand is manifested in the whimsical combination of a trailer with an old garden bench pulled by a modern two-wheel tractor.

general, they reinvested as much as possible, postponing consumption until after they had made the basic investments. However, some expected to be able to begin to take out some money in the coming year. The sample of 212 large (i.e., individually owned farms) mentioned earlier (Top Agrar Spezial, February 19, 1994, 10–13) had made average investments of 218,132 DM in 1992–93 and had an average total capital of 1,204,876 DM, of which 53,6% represented their own capital.[6] The investment our informants had to make in order to start or relaunch their farms was quite variable, depending on whether they already owned some farm buildings or had to start from scratch and also whether they planned to raise cattle as well as grow crops. One farmer with a 400-hectare farm, invested over 300,000 DM in machinery and 150,000 DM in a riding hall. In contrast, another farmer with 250 hectares of land (which he planned to increase to 300), who planned to raise cattle in the near future, expected to spend a total of some 1 million deutsche marks. In other words, the farmer who wished to engage in mixed agriculture had to be able to raise twice as much money.

Sometimes a farmer could luck out. In one case, a member of a small former LPG spoke about how difficult it had been to acquire a small barn where he had been storing machinery from a former LPG. When the LPG in question was dissolved, an assessor came to evaluate the worth of the building. He estimated that it was worth 650,000 DM. According to our informant, the estimate was so high because he would have received 10% of the sum. As we have seen, learning the "true" value of a commodity was a very complex and variable process. He recalled that all the members of the former collective were already counting on the money. Well, no one bought the building. After two years of renting it, he bought it for only 100,000 DM.

Except for those *Wiedereinrichter* who were able to raise cash by selling some of their land to developers, few could rely on their own resources for a substantial share of the initial capital necessary to start their farms. As we have seen in chapter 5, they could demand compensation for the assets in cash and kind that their families had brought into the collectives when they were forced to join. But such compensation rarely amounted to very much and it could often not be obtained without a struggle. As one farmer described it, "I went to the LPG where my father was a member and asked, 'What about the inventory my father brought into the collective? I am the heir.' The manager, [a woman], told me, 'You may come and get it, but in the form of animals, not cash.' I replied, 'Fine, tomorrow I am coming with a large cattle truck to fetch them.' She looked at me [in disbelief]. But I did go down there and took my cattle. I had to stand on my head to do it, but I was able to get my money out that way. I was the only one to do so. My father had brought in seven or eight cows. I got exactly the same number. And there were some thirty pigs. Most of the others wanted cash. I said, 'I don't care. I will also take animals.' I am the only one who is not quarreling about [my father's] entry contribution. Only two horses are still outstanding. She didn't have them. I had a lot of work, but I took that in stride. I am not afraid of work." Another informant was able to pick out some cows from the cooperative's herd to replace those his family had contributed when they joined the cooperative. Although he received only half the original number, his mother, who formerly worked as a milk controller, assisted him in selecting the best animals.

As we have already mentioned, a substantial proportion of the funds invested came from delayed consumption. The farmers calculated their own labor only for accounting purposes, but actually reinvested all the farm income while living off the money that the women in their households were able to earn from wages or benefits or, in two instances, from their own income from nonfarm sources. In the case of all our informants, reinvesting farm income was also facilitated by the fact that they owned their own houses and grew much of their own food.

However, for most small farmers, the most crucial component came from a variety of grants, low-interest public loans, and commercial bank loans with state-subsidized interest geared toward the rationalization and adaptation of agriculture and the protection of the environment.

Like in the case of the cooperatives, any request for funding required a detailed proposal submitted in a prescribed format which could, in practice, only be secured by hiring a western or western-trained consultant. A number of pitfalls awaited the unwary farmer. The process was unclear and costly. In the words of one victim, "If I want to become a *Wiedereinrichter* I have to submit a proposal for that. I am not allowed to write one myself. Only a certified consultant may do so. They do this on a computer and it costs 2,000 to 3,000 marks. I got taken right from the start. I received a letter from Bernburg that a consultant would come and see me. All I had to do was sign a request. We chatted and he did everything, but when I received the plan, it turned out to be only a consultancy rather than a proper *Wiedereinrichtungsplan*. It was useless for my purposes. So I went to the agency and was told that I should write my plan down by hand. Thereupon, it was transferred to the computer. The plan cost me 2,000 marks. Only after it was submitted was I able to receive the subsidies." The same informant conceded, however, that he did learn a lot from the consultancy, for it showed him how much land he would have to cultivate to produce effectively. Thus, he was told that the machinery would not be sufficiently utilized if he only cultivated

100 hectares of land. For that, he would need 200 hectares. If he cultivated 300, the picture would be even rosier.

Another farmer hired an eastern German consultant. The fact that this consultant had been imprisoned during the GDR for political reasons and that he acted as consultant to some of the other farmers in the area, including two of our other informants, inspired confidence. He was told that he should borrow as much as possible while the going was good, because obtaining bank credit was likely to become more difficult after the first wave of farm foundings. The consultant turned out to be right. Now the farmer was sorry that he did not heed his advice more closely. Our other two informants who employed the services of this same consultant were equally laudatory and marveled at the man's knowledge of agriculture as well as tax matters. They continued to go to him for advice and to do the tax accounting.

When all the conditions were met and a farmer's proposal was accepted, the state funds and loans could begin to flow, but usually only after new delays. Specifically, in 1990, a full-time farmer could obtain a grant of 20,000 DM not tied to investment; an investment subsidy of an additional 40,000 DM; a public loan of 160,000 at 1% annual interest and 3.5% amortization for reestablishing a farm, or 50,000 for modernizing a farm; and a subsidized commercial loan of up to 300,000 DM (later 400,000) for which the state paid up to 5% after one and three-quarters years with amortization beginning only after four years and a repayment period of twenty years for buildings and ten years for other investments, so that the actual interest amounted to 4% or less (Bundesministerium für Ernährung, Landwirtschaft und Forsten 1990). In addition, as indicated in the previous chapter, a farmer could receive up to 750 DM for every hectare of land enrolled in a long-term program of taking land out of agriculture and receive subsidies for extensifying production. Finally, in 1989–90, average annual subsidies allocated to farmers by the European Union amounted to 18,535 deutsche marks. However, in spite of the support the state and the European Union provided agriculture, banks were not very keen to lend to farmers, and there were often long delays before credits were approved.[7]

These delays made farmers nervous because the price of leasing land was going up due to western competition and indications that the customary rules of leasing were being abrogated, making tenure less secure. They were also afraid that further exchanges with cooperatives might not work out and that land they were trying to lease might be snapped up by western farmers.

As we have seen, by 1999, the informants we revisited had made substantial investments in technology. They also put new roofs on their houses and barns. The farmer who planned to construct a riding hall had done so, but had almost gone bankrupt in the process. However, for reasons we shall explore later, no one, including the family that would clearly have had the funds to do so, invested in livestock operations.

Division of Labor and the Employment of Outside Labor

Unlike the GDR collectives and, to a lesser extent, even modern farming cooperatives, where the division of labor was fairly rigid, the eastern individual farmers have become generalists and women were also involved in more than one aspect of the enterprise. Tasks were usually allocated to the members of the household. In general, wives did not receive a salary, but adult children and their spouses were recompensed. However, unlike in the past, spouses now did receive social security benefits when they engaged in private farming.

In addition to household labor, farmers hired temporary help in the peak season and recruited among retirees, the unemployed, and students. At least during the first years, when a farmer committed himself to take on an apprentice, he could receive subsidies to pay the wages. Only one farmer planned to hire a full-time, permanent farm hand and that was because the adult members of the extended household were engaged in ancillary activities.

While wives assisted in all aspects of running the farm, including activities usually defined as masculine, such as tractor driving, they also tried to carve out a sphere which they defined as their own. Wives often contributed their accounting skills to the enterprise, or were engaged in an ancillary enterprise, while older women were called upon to take care of grandchildren.[8] In 1999, the younger Mrs. Feldmann gave us an indication of the importance of dealing with the necessary paperwork in running the farm. Laughing triumphantly, she said, "My parents-in-law and my own parents don't understand it when I say that I achieve just as much with the stroke of a pen as when my husband plows strenuously for an hour. And when there is no rain or when the weather doesn't do its part and he goes through hell, I still earn my money. He gets very upset about that." In Mrs. Feldmann's case, the paperwork also includes investing the surplus in the stock market which earns her far more than the 2% to 2.5% return on a regular bank account. We found intricate divisions of labor, with grandmothers taking care of grandchildren, but also driving tractors, fathers and sons working in the fields, and daughters-in-law doing the accounting. In one instance, the wife and one daughter ran a restaurant that was supplied with produce from the family farm; the grandparents helped with the vegetable garden and child rearing; and the husband did much of the field work with the assistance of a hired woman, whose tasks ranged from tractor driving to weeding vegetable patches by hand. He claimed that men were not suited for weeding, a "traditional" task. Thus finding a woman who would also drive a tractor turned out to be the ideal choice. Unlike men, women, then, were expected to combine old and new tasks.

Like cooperatives, independent farmers also hired temporary help who were either retired or unemployed persons. In the latter case the question of remuneration became problematic. They would lose part or all of their unemployment benefits if they worked legally and earned more than 315 deutsche marks per month or worked more than 14.9 hours a week. That would mean that they were actually working for practically nothing. As one informant told us, "When you say that you have earned 100 DM, the state takes away 80. In principle, if they work 60 hours a month, they have worked for 20 DM (per day). That is exploitation. But if a person abides by the 14.9 hour a week limit, then he can work only three hours a day from Monday to Friday. He or she couldn't work for nine hours on Monday, for example. That means that I would have to hire four persons [to do one day's work]. Otherwise he or she would be classified as someone who earns more and would be treated differently by the unemployment office. They tell us that they would rather not be at home all the time. They would like to get out for four weeks and they get points for having tried to find employment, so that when there is a job opening somewhere they will have more of a chance. But let me make it clear that I don't pay 30 cents. What I write down on the piece of paper is another matter. The people also eat here. If the people at the unemployment office knew that I give them a bottle of cola when it's hot, lunch, and a sandwich in the morning, my workers would still be owing *me* 30 cents. To do things like that we still think too much as eastern Germans (*Ossidenken*). We are not west Germans who say, 'Okay, if they come for 30 cents that's fine with me.' Instead I say, 'You will get 5.00 deutsche marks per hour and I will write down 30 cents while I pay the rest out of my pocket.'"

In fact, our informant had once received written permission to pay workers 1,200 deutsche marks for four weeks of work only to be told at the end that the law had changed and that the workers could keep only 120. "Then they say that they won't come again. If someone looks for work and wants to abide by the law, he or she is penalized. It is better for them to say, 'I am taking my unemployment money and that's it.' In addition, if they have a low income, they get a supplement for rent, child support for schools and kindergartens, and 30 marks for television dues.

So if they come here they have to change all that too. And such changes can't be made just for one month. They make the change for the entire year. It is far too much trouble to make it worthwhile. That is what makes our situation so difficult. We have people who have been coming for five or six years in a row. They say that it is good to work together with others for four weeks; have some fun; hear and see something different. They don't care if the work is dirty and they are covered with dust, and they are eager to come again the next year. However, we are always in prison with one foot. In construction, they prefer to pay someone overtime or hire someone who is moonlighting for 15 deutsche marks per hour rather than risk hiring an unemployed worker. That's the way it is. Those who have work, work from six in the morning till nine in the evening, and those who don't have a job don't work at all. Also, at 42, people are considered to be too old and are no longer hired because untrained workers are cheaper. They only cost 15 marks, while a trained worker would have to be paid 8 marks more. I often wonder how people in Japan can work in the same job all their lives. Here many change jobs as frequently as they change their shirt. And many don't work in the fields in which they were trained. Of the 200 students at the university, only ten are in agriculture. If they worked for an insurance company that deals with farmers, that might still be alright, but many work in car dealerships and things like that."

Although the frugal management of an independent farm entailed the use of every free family hand and the occasional use of cheap temporary outside labor, the farmers in our sample also hired more specialized help because they lacked inputs or knowledge. Some did not own all the equipment for certain tasks. One felt that he lacked the expertise for planting and harvesting sunflowers and hired the neighboring cooperative to do the work. Others also all hired experts in enterprise planning and tax consultants.

Between 1994 and 1999, the informants we revisited had not made substantial changes in the division of labor within the family, although some did hire more outside help. The son of one of our informants had finished his studies, while the son of another one had postponed studying as long as he could and was now finally going to class in winter. As his father said, he would not have gone if he had not been forced and was surprised that one had to know things like the plans for agriculture of the European Union.

The Feldmanns had made the most changes. The older generation was slowing down (the elder Mr. Feldmann is close to 70) so they hired a tractor driver on a full-time basis and a woman who helps prepare the potatoes for direct marketing and other things like that for five hours a day. In addition, they are training an apprentice. He had come to them after quitting his job with a "flatbed truck farmer," a man who directed farming operations from the west and expected the young man to be able to follow orders by phone, although he had had no technical training at all. When it rained in Bernburg and the young man told his boss that he couldn't go out to work in the fields, the boss would tell him that it was not raining where he was and that he should not be so lazy. The Feldmanns also hire part-time workers for such tasks as harvesting potatoes. Finally, they take advantage of the availability of a skilled retired mechanic, a neighbor, to repair equipment. They compare his all-round talents to the *Heinzelmännchen*, the elves in the fairy tale who brought comfort to the townsmen of Cologne by doing all their work at night while they slept. They also enjoy a reciprocal relationship with a shepherd whom they assist and from whom they receive mutton.

In order to provide their permanent workers with retirement benefits beyond those provided by the state and, at the same time, tie their employees more closely to the farm, the Feldmanns are resorting to an unorthodox, personalistic strategy. They have established a pension fund for them which can be abrogated should there be a parting of ways within a period of less than ten years.

As we have seen earlier, independent farmers often take on additional field work for part-time farmers, who in turn assist them during harvest time or, in one case, more regularly. In one case six men help. As our informant said, "It presents no problem because I have always done this in the past as well. I helped them and they helped me." For example, there is an unemployed man with 4 to 5 hectares of land whose wife works in a factory. "He is a willing young man who has always worked in agriculture, but no one is hiring him. When he comes he doesn't think about being paid right away. He is not interested in that. But we write down what he does for me and what I do for him and he gets some money if he has worked a few hours for me and the other way around too." The same farmer also has an arrangement with another part-time farmer, a former LPG brigade leader, who works in a factory but cultivates 10 hectares on the side. "Then there is a young man who does own a combine and two old tractors. He has a full-time job in the factory but also works the land. He has a woman who drives the tractor. They are not married, but who marries these days? He makes a lot of money. He hasn't made any investments. They earn more than we large ones, because we have invested so much. He has old machines and is handy. He fixes them himself. So he earns something. Not too much either, but more per hectare than we do." Finally, he borrows equipment from a western German farm manager on occasion.

Crop Production and Its Link to Market Demand

As in the case of the cooperatives, the decision to plant a particular crop was intimately linked to the market and to the degree to which a particular crop received subsidies during a given year. It was also determined by ecological considerations including soil and climatic conditions, crop sequences, and the particular characteristics of the fields to be planted. As one *Wiedereinrichter* remarked, "We have to orient ourselves according to the market. At present, we really have to have our ears to the ground. A lot of things are distorted by the system of subsidies. We receive close to a third [of our revenue] from the state. For example, we are receiving an equalization payment on the first of December. It does not eliminate all the risk, but it is crucial to take it into account. Rape seed receives a higher subsidy than grain. So, if it makes any sense at all in terms of crop rotation, I prefer to plant more rape. On the other hand, I can't follow that path in the long run because I also grow sugar beets and I do have to pay some attention to the appropriate crop sequence. I have to fallow the land because of pests and everything. With sunflower seeds, I make contracts before I plant. I have decided to try that and see how it works."

The choice of what to grow is also strongly determined by the labor costs involved in producing a particular crop, for, unlike the cooperatives, individual farmers work with few outside workers if any and any increase in labor tends to come out of their own hide. In particular, the independent farmers must pay careful attention to the specific seasonal labor requirements of each crop, so that the work can be spread out as much as possible. Also, most *Wiedereinrichter* shun capital-intensive specialty crops such as medicinal herbs. Even a group of farmers who once specialized in growing herbs no longer grow them. Finally, *Wiedereinrichter* are somewhat more likely than cooperatives to gear their production toward local markets.

As in the case of cooperatives, a high percentage of the production was in wheat, rape seed, and sugar beets. They also grew legumes such as peas and fodder crops such as lupines. Some, like the farmer in the example above, have experimented with sunflowers for which there was a rising demand, although growing conditions in the area do not appear to be ideal for that crop. Farmers were more tentative about planting barley for brewing even though it was once common in the region. Although it fetched high prices, the high industry standard increased the risk; for if its protein content was too high, it had to be sold as fodder barley that fetched a lower price than if the land had been planted in wheat.

Like the cooperatives, the individual farmers were affected by the volatility in demand resulting from the eastern German consumers' experimentation with new foods from the west. One *Wiedereinrichter* complained that "We grew a few denser varieties of potatoes [of the kind preferred in western Germany] because some of my clients told me that they wanted firm ones, claiming that they were eating only that kind of potato [now]. But this year, they told me that they really didn't like them after all. [For] in this region people [generally] say, 'Just keep those dense potatoes away from me. I want the mealier ones.' [With the potatoes], it was the same way as it was in the beginning with the buns. At first, everybody wanted the western type. Now they want to have the nice, hearty buns from the baker again. At first, people tried everything else and then they returned to the old things. [The mealier varieties of potatoes] are used to make mashed potatoes, for which many other varieties are not suited. We live with the risk, of course. When I plant [a particular] variety and then have a thousand hundred weights of potatoes [that no one wants], what do I do then?" The same farm family had similar problems with the size of onions. In this case, the size and shape turned out not to meet the current fad in that particular year. EU norms standardize and constrain production and do not take local culinary traditions into account. Finally, the fiasco resulting from the inability of the monopolizing sugar factory to keep up with the bumper crop produced during the year of our study, led some of our informants to reconsider the amount of land they planned to dedicate to this mainstay of the region.

A favorable location and strong ties in the area enabled some *Wiedereinrichter* to combine production with the retailing of their produce in order to carve out special market niches for themselves in the area. The Feldmanns, whose farmstead is located in the middle of a small city, took advantage of its location to supply the town with potatoes. Mr. Feldmann explained: "We started early. Before, potatoes were never grown in this area, in spite of the good soil. Even today, people disapprove of growing potatoes. I can say with pride, that we are supplying the entire town with potatoes for basement storage. That didn't exist before. The potatoes had to be trucked in from many kilometers away. We are diversifying our crop sequence in this manner at least on those 3 hectares."

Thus, for the Feldmanns, provisioning the local market worked very well. Other alternatives would have been more difficult: the larger industries demanded potatoes of a particular size because their machines were only equipped to deal with a narrow size range and smaller buyers usually already had their established sources to whom they gave preference during years when there was a glut in the market. As Mr. Feldmann remarked, "We tried it out and took the risk. We stored the potatoes in our barn and warmed them with blankets, tarps, and hot air when the weather was too cold. And until now, touch wood, it has worked out well. We only have a few left and we have to make sure that we have enough for the kindergarten and the old age home that we always supply with potatoes. They will get some until the new harvest comes in. They are our best clients and come every week. This kind of steady client makes us realize that [our efforts] have not been in vain."

Another farmer has decided to raise vegetables, although he does not consider himself a horticulturalist. As he claims, "I have the vegetable field down in the wet meadowland near the Saale River. There, I have enough water so that I can irrigate. The land is capricious. It can be very wet, but also very dry which makes it difficult to work. If I plant things that produce above the ground, I have fewer problems. I raise the locally traditional crops, especially [various varieties of] cabbage." There is little competition in this area. He attributes his success, in part, to the cooperative arrangement he has been able to forge with a horticulturalist. "I produce the vegetables and my gardener colleague, who has three market stands in Bernburg, sells them along with his own [which he obtains from other sources]. . . . He grows cherries on several hectares of land and some vegetables, but only on the side, and plans to grow apples. We try not to compete with one another. I have a lot of irrigation water and he has only a little. We also cooperate in that I truck his produce to the market. A lot of empty crates are always left over from his business and I can use these for my vegetables. They save me about 20%. I hope that I can adjust my production better to the market next year. At present, I only plant 2 hectares, which is a small part of my holdings. If I hire another farmhand, we will have more time for the vegetables." His observations of produce delivered from Holland has also taught him about the value of modern means of ensuring freshness. "Last winter I already refurbished a trailer with refrigeration. When I harvest cauliflower, I can refrigerate it with a generator. In this manner, I can reduce the amount of chemicals [e.g., sulfite sprays] that I would otherwise need to maintain freshness. That doesn't mean, however, that I am an organic farmer. That, one can only practice as a hobby."[9]

Specialization based on particularly favorable location and ties to a particular clientele are not the only ways that a farmer can have an edge over his competitors. The judicious application of specialized skills acquired during GDR times can have a significant general effect on the viability of the farm. Thus one farmer saved considerable amounts of money by frugal use of pesticides. To the amusement of his neighbors, he would spray in the evening, when the sun was no longer hot and there was little wind, because this was best for the plants and reduced the amount of chemicals required. Another farmer calculated nitrogen fertilizer very carefully. He figured that nitrogen had not been used up because of drought conditions in the previous year; he could save by waiting for another year before applying nitrogen again. As in personal consumption, frugality in the use of productive inputs was regarded as crucial for the survival of the enterprise.

Astute farmers could obtain advice on the use of these inputs for free if they chose their suppliers carefully. As one informant explained, "The dealers who want to sell us pesticides also act as consultants in order to improve their standing in the eyes of the customers, so that we say, 'This man doesn't just care about selling, but advises us what to use in a specific location.' After three or four years, each farmer has come to choose a firm from which he makes his purchases. So the dealer will come out and look at a field and say, 'Last year we had bad luck with this product, so why don't you try that one this time.'. . . In the beginning there were all sorts of suppliers, some of whom had been engaged as specialists on farms, like my father-in-law. Some of these were friends or acquaintances. In the past three years, the grain has been separated from the chaff. We may go to someone with whom we were not acquainted before because he is better." Nevertheless, this farmer preferred not to buy from dealers who came from far away because when there was a crisis, a specialist had to be able to come right away.

As in the case of the cooperatives, between 1994 and 1999, the eastern independent farmers have undertaken few experiments with new crops beyond growing linseed. One family stopped even that because it meant taking out their machines once more to harvest the linseed after all the rest of the crops had already been harvested. "Without this," said our informant, "one is finished for the year, can clean the combine, and can store it away." And the individual farmers have stopped growing sunflowers too because the growing conditions were less than optimal. Mr. Feldmann commented on the farmers' reluctance to introduce more variation in their crop mix and on the resulting monotony of the landscape. "There is this wonderful euphemism here 'appropriate farming practice' ('gute fachliche Praxis') which really means to plant something that doesn't bring much profit. But it is true, nature is revived when there is one type of crop standing in one field and another right next to it. When I drive from Halle to Magdeburg, all I see left and right of the highway is wheat, sugar beets, and rape, nothing else."

Mrs. Feldmann felt that the crop mix would become even narrower if the EU equalized its subsidies for all types of crops as it was planning to do within the next few years. For then, only those crops for which the region is best suited could be grown profitably. Nevertheless, the Feldmanns continue to grow potatoes for direct resale, spraying them with as little pesticides as possible to avoid allergic reactions in their more sensitive customers, with the help of the woman they have hired. An additional woman is hired on a temporary basis to plant small amounts of beets, red beets, carrots, and fodder beets—a highly labor-intensive activity, since all the transplanting and weeding has to be done by hand—which they also sell directly to consumers, and they are cultivating durum wheat (used for noodles) which is normally grown only in more southern climates but appears to do well in the dry Harz region.

Livestock Production

Unlike the cooperatives, which continued to raise livestock, although in reduced numbers, the individual farmers (and as we shall see later, particularly those from the west) have tended to concentrate more heavily on crop production. This tendency was particularly salient among the larger farms, while the smaller ones placed more emphasis on livestock production (Top Agrar Spezial, February 19, 1994, 11). Out of the six eastern German *Wiedereinrichter* we interviewed, only one was raising cattle in 1994. He had thirty cows, but he had sufficient milking equipment and a high enough milk quota to keep an additional fifty as soon as he could enlarge the barn. Another was raising some eighty pigs. He had given up raising cattle, since his sons showed little interest in raising animals. Two of our informants, both of whom raised animals before the *Wende*, were, as we have seen earlier, limited in the expansion of their operations because their barns were located within city perimeters. One of them was planning to build facilities for raising bulls and keeping milk cows. In the meantime, he and his family were raising a few pigs and keeping a horse or two in their barn located in the middle of the city. The other would like to do so eventually if he could find a way to relocate outside the city. As things were, he had limited himself to raising a few pigs for home consumption with the potatoes left over from his potato-retailing operation.

Another farmer was renting out space in his family's barn and providing fodder to horse owners from the nearby city. His grandfather continued to own two draft horses and the daughter owned a horse as well. As we have seen earlier, he also built a covered riding rink. Because the cows which they kept in a field some distance away kept escaping, the family finally gave up raising cattle. However, they planned to fatten bulls as well as produce milk again on a larger scale in the future. Indeed, in spite of the problems involved in setting up a viable animal-raising operation, only one farmer in our sample, whose speciality—in addition to the usual crops—was truck farming, had no plans to raise animals eventually.

The dilemma faced by the farmers was not so much whether or not it made sense to raise livestock, but whether they could achieve the large scale that was required for livestock farming. Although some farmers continued to engage in small-scale animal raising, such activities, if they went beyond the provisioning of the household, were becoming increasingly unattractive.[10] One cooperative manager estimated that out of the fifty members of the cooperative fifteen still raised pigs; before, the same members raised a total of some fifty to sixty pigs. As one farmer explained, "The market has to be there. A family to whom we sell potatoes has three cows. But the problems begin right there. Someone has to take care of the calves. If one has a cow that produces 30 liters of milk a day, where am I going to go with all that milk? One has to have private customers. Then, sooner or later, the hygiene authorities will appear on the scene to check the way the milk is being handled. So I would say that one has to start out on a larger scale. Then there is the problem of the milk quota. Each *Kreis* has a certain quota, so if I want to produce milk, the LPG (sic) has to cede part of its quota to me." Another reason for giving up small-scale production was the fact that there was less economic pressure to engage in supplementary activities. When we asked one farmer whether the fact that prices for such products as eggs were no longer subsidized had affected small-scale animal raising, he answered, "At the time we received 30 pfennig for each egg. Well, at present we are again receiving 25 or even 30 in the city. But one has to look at it from another side. People used to earn less and so they needed a second income. Everybody had to engage in some supplementary activity and that was easiest with eggs. I would just have to throw out some grains and let the hens run outside. Only the LPG still raises broilers. The others just hold ten or fifteen for their own consumption. We still have eighty. It's the same thing with fattening pigs. It's not worth bringing them to the slaughterhouse. I don't get a decent price. I would have to raise a thousand right away. But I still raise some. We still butcher some ourselves, and acquaintances come from our village who prefer to make their own sausage. . . . The butchers buy their meat at the slaughterhouse, where it is cheaper. But the home butcher doesn't mind paying more for a pig from a known source. It's the same thing with hens. Lots of people come from Bernburg to buy eggs from me because they know that they run free and are raised properly."

As before the *Wende*, animal raising as a subsidiary activity continued to depend on recycling leftovers from human consumption. The reputation of the GDR for wasting bread not withstanding, persons who continued to raise pigs appeared to have no difficulty in obtaining large quantities of day-old bread today as well. As one farmer said, "Before, I did not get as much bread [for my animals] as now. Today, I still have a baker where I get more bread than before. Today the clients are even more picky than they used to be. What they bake today, they can't sell tomorrow. In the GDR, they were allowed to do so. [But it is true that in the past] bread was fed to animals by the ton. After all, it cost only 1 mark. So I would get what the people threw away. Next to the apartment houses there were special cans for feed. Enormous feed lots lived from that. I also had my sources. Every week we would drive around and fetch it. Today, that is no longer allowed. It has to be composted. The bread from the baker is a different matter. It is untainted. [That too I get for free]. I could feed the animals grain. But I have gotten used to doing it this way."

Three or four shepherds have reestablished small herds of 300 ewes or so, mostly on land suitable only for pasture. Our informants, cooperative managers who provide these shepherds with land and sell them fodder and straw in return for dung and individual farmers who sell surplus sugar beets to them concur with the author of an article in *Top Agrar Spezial* (Grund 1991:6–10) that because of the low prices for wool and mutton, herding sheep permits a marginal existence at best. Thus one shepherd who had worked for an LPG with a herd of high-quality, registered sheep, keeps sheep purely for meat; shearing would cost more than the wool produced. The venture is profitable only if he keeps the sheep until they weigh 35 kilos. In addition to grazing the sheep, he has to feed them special fodder to bring them up to that weight as rapidly as possible. Since the thruway will cut across the land he has rented to pasture the sheep, he will no longer be able to pasture them on the other side of the road. Even now three persons have to block the highway to herd the animals across. The loss of land occupied by the thruway may also induce the authorities to force him to reduce his herd. Originally, he worked with his daughter but she had to find another job because they could not make ends meet. In two cases, sheep herding is complemented by family income from retirement and unemployment benefits. In one of these instances, the retired father tends the herd, which enables the son to continue to draw unemployment benefits. It seems that, at present, only a small family with at least another small source of income can manage to subsist on sheep raising. Our informants lamented that during GDR times every village along the rivers

FIGURE 7.6. *Wiedereinrichter* often raise a few animals more for the sake of nostalgia than as an economic venture.

kept a herd of sheep to keep the slopes stamped down and clear of weeds. Now the government subsidizes herds to this effect, but there are far fewer sheep now which has already resulted in rivers breaching earthen dams that have become too soft. In some areas shepherds are receiving payments for keeping marginal land clear, a practice that may well increase in time.

As we have already indicated earlier, those farmers who had not already begun to raise livestock on a larger scale did not expand their operations much less build new facilities by 1999. Indeed, it was doubtful whether animal husbandry is still at all economical if the farmer's labor is figured as a cost. Why farmers continued to bother raising some animals at all made some sense to us when we thought about the cozy, cheerful scene in the barn that the family who had moved from their farm in the middle of a city block to the outskirts of Bernburg. In a wonderful old barn, they kept an old horse, several dwarf goats, a few pigs, and a calf that was being fed by the part-time, retired worker as we watched. "We also still raise some chickens which we already had in Bernburg," explained the son, "Italians, and a white cock. We only have seven. Maybe we will have more later. We didn't have a cock in Bernburg. He would have crowed." "Yes," added the mother, "I beat him down with a stick and told him to stay there, but he pulled himself together and got up again." The family was obviously fond of these "pets." However, after already having cleared an area of ruins and sorting the debris for disposal according to regulations for a barn to raise beef cattle, this family also abandoned the plan as uneconomical.

Prospects

From the foregoing discussion, it should be clear why relatively few individuals decided to establish an independent farm. Although some of the risk was assumed by the state and the European Union, the individual farmer was still required to have the necessary credentials and managerial and practical expertise, as well as practice a very cost-conscious and market-oriented agriculture to survive. He and his family were also required to continue a modest lifestyle and postpone consumption and leisure. All of them said that they were certainly better off than before and, in spite of their hardships, positive about their life choices.

Indeed, by 1999, the farmers' sacrifices appeared to have paid off quite handsomely, judging by their lifestyle and investment practices. They enjoyed new appliances and other home furnishings. They were taking vacations in other parts of Europe, buying powerful cars, and providing children and grandchildren with "luxuries" such as a horse and a horse trailer and trips. Invited by the John Deere company, the elder Feldmanns took a trip to the United States themselves, for, afraid of an airplane crash, the younger couple did not want to fly unless the entire family was along.

The possibility, both financial and political, to travel abroad, as well as the other opportunities for international communication have widened the horizon of our informants. As the younger Mrs. Feldmann explained, "Last year we were in Holland. Two years ago we were in England. We looked at farms there. We do get interested and we would like to get to know people. After all, today we are allowed to travel. Until 1990, we had been impeded from doing so. And [in our travels] we always have in the back of our mind the prospect that our daughter will soon spend some time on the other side of the Big Pond (the Atlantic). . . . My parents-in-law were delighted [with their trip]. It was an organized tour to Canada, America. They were away for eight days. They came back with seven or eight rolls of photos. They also telephoned us. One has to distance oneself from one's work. Otherwise one becomes blind to the world outside of one's work (*betriebsblind*). One becomes irritated about every little thing."

Another family had already taken part in an exchange program with a family in the United States. Our informant, Mrs. Z., was apprehensive. When her daughter called she had felt, "Oh God, she is so far away, all alone. If something should happen, we wouldn't even know where to look for her." She was nervous, but also delighted to be able to communicate over the phone with her daughter's American host. "At first I was shy and couldn't understand what he was saying, but then he said, 'S. is a very nice girl.' I understood that immediately." A group of high school pupils and the mayor of the American town had visited Bernburg. The (American) teenagers were delighted with the fact that they could roam around freely. "They don't do that at home. They are much more strictly supervised. It must be because of crime." But money is also spent in other ways. On the other hand, the family that struck gold because of the urban land they owned (and still own) is quite happy to remain in Bernburg and enjoy their new house. As Mrs. X. saw it, "We don't drive anywhere. Where should we drive to?" One of her two sons, both bachelors, added, "We did go to the farm equipment fair in Hannover. But we have peace and quiet here, so we don't travel around." "There is also the desire to make things nice around the house," continued Mrs. X. "That costs money. We have beautiful weather here. So we sit outside and drink a cup of coffee or eat our evening meal."

The increasing prosperity of the farmers is beginning to have repercussions on their relationships with less fortunate members of their communities. At this point, class differentiation is still muted. Eastern German cooperative managers and private farmers often still live below their means compared to those western farmers who have already built or refurbished new homes in the east. However, wealth differentials are already having some adverse effects. A former manager of a tractor pool of a large state farm who now holds a high managerial position in a tractor distributorship lamented the divisive influence of differences in wealth. When we asked him how envy expressed itself, he answered, "People no longer talk to one another as openly as before. People still greet each other, but it doesn't go much farther than that. Before, we might sit in the garden together, chat, and drink beer. There is less and less of that. . . . Or, before, everybody knew exactly how much everybody else was earning and spending. That is impossible today. Before, when decisions had to be made regarding wages I would call all thirty workers who worked in the shop together and we would argue how much each one would receive in wages and premiums. We would discuss why this person would receive 10 marks more, why that person would receive a premium of 960 marks and the other 980. This would go on for half an hour with all thirty present. One couldn't even think of doing that today! In these matters, people were open and honest. One also knew that the director or manager was earning 500 marks more, at the most. For that, he had to work many hours and was not paid for overtime. In contrast, a worker who sat in a threshing machine and accumulated 100 hours of overtime during the summer earned more during that time than the manager, even though the latter had to be the first to be present in the morning and the last to leave in the evening. And, at the end of the year, the monthly average was the same, because those who worked fewer hours would do something at home such as raise chickens or rabbits where they could earn some extra cash. One could do whatever one wanted. . . . Everybody had the same income. Those who wanted to build a house did so whether they were managers or stokers. Those who wished to do something after working hours did so. Or they built a bungalow or bought themselves a motor boat or engaged in some expensive sport, within what was possible at the time. . . . Now differences are beginning to manifest themselves. Those who have two children and where only one member of the family works are feeling more pinched, while where two are working and are earning good salaries, and where the children are, perhaps, already out of the house, the couple can invest and do something. . . . One can already see how differences are emerging. . . . For example, some still heat with coal or wood, others have oil or gas heating systems."

While he himself still seeks out his former colleagues and friends irrespective of differences in wealth, this informant feels that his less successful friends are becoming more reserved. "That circle is shrinking. People observe what one owns or does. They will ask where you are going for a vacation and if you say that you are taking a trip to the Harz Mountains they will say, 'Oh, I see,' but if you say that you are going to America then they will ask, 'How can you afford to do that?' and so forth. On the one hand arrogance and on the other hand envy have appeared." Berdahl (1999:135) has made similar observations regarding emerging social stratification. Informants with jobs found that they could no longer "sit together" with other villagers who were unemployed or who earned much less because they could not stand their envy.

Emerging class differentials may have even more serious consequences. For example, in one case, the grandchild of a farmer was subjected to extortion by schoolmates. Envy may well increase further when the farmers have absorbed the start-up costs of their operations and are in a position to show their wealth through greater consumption. Indeed, the newly independent eastern farmers have been postponing consumption in order to be able to invest as much as possible into the farm. They are used to living more frugally from GDR times and criticize their western counterparts for living beyond their means. Easterners are also concerned about the repercussions of unequal access to education and other services, which may further exacerbate socioeconomic differentiation. These considerations point to the very real possibility that much of the pain of increasing economic polarization may yet have to be borne.

In spite of the present relative prosperity, the independent farmers, like their cooperative manager counterparts, are worried about the future. Mr. Feldmann, junior, answered our question about the future of agriculture as follows: "[I see the future] with mixed feelings. [According to the new plans] there will only be crop subsidies until 2005 or 2006. Sugar production would no longer be regulated and there would no longer be any milk quotas, so that the market would regulate everything. But if the first hectare gives a negative balance, how can the thousandths make a profit? The economists have found that after 750 hectares there are no further economies of scale. The largest cost digressions are between 250 and 500 hectares. The ratio of cost to profit has gotten worse in recent years. Yields have decreased and while costs have decreased as well, they have not decreased as much. The Agenda 2000 (the new economic plan) proposes reducing subsidies by 100 marks per hectare. The maximum income allowable [to still receive the supports] are to be reduced. The amount that is free of taxes is to be halved. Two years ago we paid one mark for a liter of diesel. Now we pay 1.40. I need 80 liters per hectare. Now they want to either sell diesel for the same price as gasoline or place a tax on agricultural vehicles. . . . More energy-efficient tractors won't make up for the increase in energy costs. For example, a machine that doubles the work of an old one is 3.5 times as expensive. The optimists say that if you can't engage in agriculture profitably between Halle and Magdeburg, you can't do it anywhere. Some say that everything will have to be mechanized so that you can do everything yourself. It would only make sense to have cattle if you already owned a barn. For one has to figure 80,000 marks per cow." He figures that he could still increase the size of the farm, but then he would have to rent land farther away, adding to traveling time, and he would have to buy larger tractors.

Conclusion

Contrary to what one might expect, most of the eastern German independent farmers had less of a problem constructing their identity than their cooperative manager counterparts. They felt that they had just picked up where they or their families had left off in the 1960s. In addition, some could point to the fact that they had never totally abandoned private farming during the GDR and many of them were either unwilling members of the cooperatives or occupied positions that gave them a degree of autonomy. The battles with the cooperatives to gain access to land made them relive the forced inclusion into the collectives when they were formed, further strengthening their sense of historical distinctiveness. A notable exception is the cooperative manager turned demesne farmer who bridges this gap by identifying with both camps. He retains close ties with the cooperative at the same time that he is beginning to identify with other demesne farmers, including westerners in the wider region.

The *Wiedereinrichter* were secure in their historic claim to place and farming tradition. Unlike the westerners, many could point to local roots, and those who had established themselves in the region before collectivization had run its course could point to their entrepreneurial spirit, which had induced them to take over a farm in the area at a time when others were abandoning individual farming. Indeed, many did not stress rootedness specifically, contenting themselves with showing us the physical evidence of this rootedness: the barns, homes, courtyards, old horses, and old equipment. The German word *Hof* (literally, compound) includes houses, barns, and a claim to stability. This security could now be enhanced by developing local ties even further, in contrast to the cooperatives, who were forced to disengage themselves from communal obligations. The *Wiedereinrichter* became active in voluntary associations, headed town councils, and provided locally defined services by engaging in reciprocal work relationships with part-time farmers and selling directly to local consumers whose earlier fickleness had been replaced by the nostalgia for regional specialties. Their security was further enhanced by the fact that, at least officially, independent medium-scale farming was the preferred option of the government, manifested in policies regarding access to land, subsidies, and credit, although, in actual fact, many of these policies were not honored.

In practice, the behavioral changes associated with individual farming have been more pronounced than the *Wiedereinrichter* perhaps realized. The members of the extended families were dependent on one another in ways that they had not been before. For example, when one of our informants suffered a brain hemorrhage, one of his daughters, who runs a restaurant, took over the daily management of the farm. And, by 1999, the farmers were increasingly scrutinizing the skills and predilections of their young children and grandchildren for signs of interest to assure the continuation of the farm. This attitude represented a change from 1994, when only older farmers sought to secure successors by training one or more sons for the task. One factor in this change was certainly the fact that the memory of secure employment was fading. Another was the fact that the farms were now well established and represented a hefty investment in funds and energy which the founders did not wish to dissipate. Potential successors now included both boys and girls.

While the *Wiedereinrichter* fit the hegemonic western German model of the family farm and *mittelständische* (bourgeois) enterprise with its stress on familism and individual entrepreneurial energy, the neo-liberal tendencies in the economy are already favoring larger enterprises, a fact that these farmers recognize and try to achieve by renting more land, even if it is located at a greater distance. They fear that their somewhat privileged position will be at risk with the prospective dismantling of agricultural subsidies and the further opening of international markets. As we shall see, the discrepancy between the ideal of the family farm and the need to achieve ever greater economies of scale is even more salient in the case of the western farmers who have established themselves in the region, who are the subject of the following chapter.

CHAPTER EIGHT

The Western Farmers

"Shall we define ourselves as migrants or intruders? The people in the east say that we are intruders. 'Immigrants' would be a more positive term, but 'intruders' . . . ? When one feels pushed into a corner—and the LPGs (sic) must certainly feel that—then one doesn't have such a positive attitude toward the intruder, even though we do already talk to each other over the telephone frequently. But that doesn't lead to friendships yet. That would include inviting a neighbor for beer in the evening." The foregoing quote of a farmer from western Germany who settled in the east captures the tense relationship between the newcomers from western Germany and beyond and the native farmers. This unease is not surprising given the differences that characterize the groups. Even the term *Neueinrichter* (newly established farmer), used mainly by the eastern farmers, is contested, entailing, as it does, the notion that this group has never had any ties to the region, but is establishing itself anew. The farmers who came from western Germany often considered themselves to be the successors of family members who had once farmed in the area and who had fled the excesses of communism and therefore as *Wiedereinrichter* like the eastern independent farmers whose families once operated farms. The fact is that former farm owners from western Germany leased 11% of the land in Sachsen-Anhalt, while western Germans with no such connections leased only 7% (Luft 1997:40).[1]

Given the extent of the differences in background, scale of farming operations, and commitment to the local community, it is indeed difficult to regard the westerners as a group at all. Indeed, for everyone, including the anthropologists, the term *Neueinrichter* did not include the *Tiefladerbauer* (flatbed truck farmer) who cultivates the land mainly in absentia, appearing only for major agricultural tasks with his equipment loaded on a flatbed truck and hiring resident labor for tasks in between. All resident farmers, even those who often traveled back to western Germany themselves, regarded such farmers with abhorrence, accusing them of violating the land by using up its residual fertility. Here we follow the eastern German categorization that separates *Neueinrichter* (mostly westerners) who did not own land in the region at the time of the *Wende* from *Wiedereinrichter* (mostly easterners) who did and from *Tiefladerbauern* (invariably westerners), although they could all be lumped together as private farmers in opposition to cooperatives.

The backgrounds of the western farmers we interviewed were more varied than those of the easterners, as were the roles these farmers played in their localities of origin. Unlike the east, with its large-scale collective farms, small- and medium-scale family farming continues to predominate in the west, even though there, as elsewhere in Europe, the tendency toward concentration into larger holdings is also progressing. In the early 1990s, the average size of farms run by full-time farmers in the Free Republic of Germany was still only 29 hectares. In addition, there was an equal number of much smaller farms operated on a part-time basis (Schunter-Kleemann 1995). Because of their small size, farms in the west tended to be more specialized than the eastern collectives in order to achieve greater economies of scale, even though, unlike in the GDR, livestock cooperatives produced much of their own feed (Gollnick et al. 1990:60).

Our western informants included six western Germans and one Dutch farmer. The western Germans included four owner-operators, one of whom also hired himself out as a contract worker for another western farmer and for a mining operation; one hired farm manager; and one owner-manager who also managed the farm of another westerner. The farmer from Holland also owned and operated a contract farming operation. They ranged in age from their early 20s to early 60s. All had other, and more attractive, options of securing a livelihood than their *Wiedereinrichter* counterparts. Indeed, it was these options that provided them with a safety net to embark on a venture in the east that entailed some risk. One farmer would even have had the option of taking over the family farm in the west, but was looking for greener pastures instead, while others felt that their only other viable options would have been to work for others.

Out of the seven, four of the owners had ties to the area and three had no ties at all. Three of the farm operators with ties to the area had taken advantage of the opportunity given to former large estate owners, often nobility, to exercise the option of renting land, presently held by the state, that their families had once owned.[2] The most straightforward of these cases was Mr. R., who explained his decision to establish a farm in the area as follows. "The farm in Q once belonged to my [maternal] grandfather. It was an old family concern [a farm of 500 hectares]. Like all farms over 100 hectares, it was expropriated in 1945 and my grandfather had to leave Q. After German reunification, we had hopes that the expropriation would be annulled. But this was impeded by shrewd maneuvers and treaties. The reunification agreement contains a clause stating that whatever had occurred during the period of occupation (i.e., under Soviet law) could not be undone. The claim was made that the Soviets had insisted on this. Now we know that this does not correspond to the truth. Rather, the intent was to use the proceeds of the sale of this real estate to finance reunification. This realization has embittered us. It violates constitutional rights. It wasn't so bad for our generation, because in our minds, we had given up these properties in the east. But for my mother who was born here and for my brothers who were older [the estate] still meant property and home. The older generation suffered a lot [from this decision]. . . . Worlds collapsed, and I thought I would have to continue as a farmer in [my home *Land*]. But then, it turned out that the *Treuhand* was willing to give preference to former owners in leasing the land, and under very good conditions. We were given four weeks to make a decision and find a bank. We began here on July 1. With my 30 years of age, I was very brave, perhaps even foolhardy. In retrospect, we are glad that everything worked out as well as it has."

In addition to 500 hectares of his grandfather's former land, he was able to rent 450 hectares from another noble family who had obtained access to the land in a similar fashion. Since all this land had been cultivated by a state farm, there were no conflicting claims.

An analogous, but more complicated case was that of Mr. von L. Although he himself was not a farmer, he decided to lay claim to his mother's land and, with his brother as partner, farm it with a certified western farmer. As in the previous case, he was able to include the land claimed by another former estate owner who lives in the United States as well as that of some relatives. Altogether the farm encompassed 730 hectares of land. Although, as the child of a conspirator against Hitler who was killed during World War II, he theoretically had a stronger claim to his mother's estate, his endless and impassioned petitions for restoration of ownership had yet to bear any fruit. The third consultant was the manager for two absentee noblemen of a farm which, for tax purposes was composed of two legally separate enterprises, but which for all administrative and operative purposes was a single unit. He formerly managed one of the three farms his actual boss owned in western Germany. After reunification, he and one of his bosses, a nobleman with three farms in the west, heard about a collective in the process of dissolution. At a meeting organized by the former managers of the collective, our consultant and his boss were able to lease 650 hectares of land from the members. Later, they heard about the plans of another nobleman to reestablish a farm on his family's former holdings that had been part of a state farm. They soon discovered that the nobleman in question was the brother of a man from whom they rented land in the west. Since neither the nobleman nor his wife had any training in agriculture, it suited them to go along with a joint project. They presented their plan to the *Treuhand* authorities in Magdeburg, who granted approval within three months. "That," he said, "would no longer be possible today. Those were excellent connections. Let me tell you, here connections are half the battle. We really discovered that here in the east. When one knows someone, then things work out and when on top of that one enjoys a good reputation, then things work out even better." All in all, the farm comprises 1,700 hectares of land, one of the largest private farms in the area.

In chapter 2, we already presented the background of the fourth western German informant, Mr. Rascher, who farmed 320 hectares in 1994 and who came to the east through contacts with an eastern German relative.

The fifth western German, with a farm of 290 hectares, had no previous ties in the area but came there after a lengthy search for a suitable farm. He explained: "I have studied agriculture and have a farming background. I have four brothers, two in forestry and two in farming. One of the latter has already rented a farm in the west, the other took over the family farm, but we couldn't find one for me. Consequently, I took a job as a consultant in a sugar factory, which I didn't like at all. After the borders opened, I hoped that I would be able to obtain land from a collective that had decided to dissolve, because I was aware of the fact that it would not be advisable to go somewhere and push someone else out. That would not have been a good start. I didn't get very far with this strategy. The land parcels escaped my grasp, and the train almost left without me. Then I petitioned the government of the *Land*. They had just announced their intention to lease their demesnes in the newspaper. I had to fill out and submit a questionnaire where I had to indicate my professional suitability and personal background as well as my financial backing. Ten farms were put up for lease and, after looking at their size, I decided on this one. I had to estimate the amount of land I could cultivate safely with my capital. We (i.e., the informant and his wife) decided that 300 hectares would be the right size. There was another one, a wonderful one, of that size here in the area. This one here is not as good because the land is prone to flooding. Although there are many buildings, they are not very attractive, and are in a bad state of repair. But we decided to bid for this one anyway because we expected the number of interested persons to be small. And we were right. Only a few petitions were submitted."

The last western German farmer in our sample, who also did not have any ties in the region, came to eastern Germany in search of land when his lease agreement with a landowner in the west had run out and the latter's son-in-law wanted to take up agriculture himself. Although he and his wife were close to retirement age, their youngest child, a son, was studying to become a farmer and so they decided to look for a suitable place to start a new farm. He told his son, "'Here (in the west) there is little land. Everything is crowded. There are a lot of young people in the village, many of whom are sons of farmers.' Then I drove around and looked for something. I figured that because of the *Wende* we might have the opportunity to enlarge. Whenever I had free time, every week or two, I came over to the east and looked, sometimes here and sometimes there. I also put ads in the paper. Finally, I had narrowed it down to three places, but we ended up here. One of the criteria was the possibility of hunting. That's my hobby. We heard about this farm through an ad in the newspaper. The owners had three large farms that they put together and took out of the LPG. They wanted to get something out of their land again and so they put in an advertisement. One of the heirs who was in southern Germany, in Bavaria, took the matter into his hands and that is why it appeared in a western newspaper. So we got in touch and we got into the final cut. There were a lot of other applicants, but we won out. We were able to purchase the homestead and we signed a twenty-four-year lease for the land." In addition to this land, 220 hectares, he was able to rent and purchase 414 hectares more from other individuals. The farmhouse itself belonged to two ladies who were well over 80 years old.

Finally, the farmer from Holland credited his move to eastern Germany to a fluke. His partner in a contract labor operation in Holland had been looking all over Europe for a farm. He had found one in France, but it didn't work out. Finally, his broker in western Germany phoned him with the news of a possibility to rent land that had become available from the dissolution of a LPG in the Bernburg area. The partner, in turn, told our informant about the opportunity, and both established separate farms, with our informant engaging exclusively in crop farming on 600 hectares of land rented from 230 individuals in addition to engaging in contract labor for others. His former partner decided to establish a dairy farm.

The information flows and social connections that brought our western informants to the east thus ranged from "modern" impersonal communication links to what remained of age-old familial class privileges to ties with specific kinsmen. These ties weathered, and paradoxically were even strengthened by, the attempts of the GDR government to limit and control interaction between east and west. Most remarkable was the reemergence of aristocracy in an area where class privileges had seemed to have disappeared long ago. The return of the influence of aristocracy was regarded with ambivalence by most parties concerned. As we have seen, German law has not recognized claims to ownership of large estates. On the other hand, by giving the former landowners preference in rental, they have created the possibility not only for major class divisions but the eventual return of the ownership through the avenue of favorable, but yet to be clarified, lease purchasing options of the rented land to the noblemen and to other wealthy investors. The government's position then was highly ambivalent. While recognizing both the influence and the possible financial role in agriculture of the noblemen, it made every effort not to be seen as supporting the return of an earlier social order. For one, the government wished to promote rationally managed farms rather than status-enhancing estates. Such an emphasis would have been particularly wasteful in Sachsen-Anhalt, with its exceptional soils. It also did not fit well with the image of a modern western state. On the other hand, the ideal of rootedness in the region combined with financial strength gave an impetus for the former landed elites to return.

Adapting to Life in the East

One of the assets that our western informants had in adapting to the east was that they did not consider themselves as ordinary western farmers. Like the easterners, the westerners regarded themselves as rebels. Their rebellion was different from both that of the managers of collectives, who had viewed themselves as resisting the encroachment of the state on the last vestiges of cooperative independence, and of the eastern independent farmers, who saw themselves as finally having won a battle with the former LPGs that began with their forced absorption. Theirs is a rebellion against the constraints of western German farming, especially the tight land market. They see themselves as pioneers in an uncharted, and yet not remote territory, a spirit of adventure that becomes apparent from the quotes in chapter 2.

Even those westerners for whom the difficulties they had experienced in getting ahead in the west were their principal reason for establishing themselves in the east stressed that they were different from their peers back home, for they were willing to forego the comforts there. When we asked Mr. Rascher, whom we have encountered in chapter 2, whether he and his wife would be able to create a life in the east, he replied, "Well, my wife is also not your ordinary westerner. She comes from a farm background and is the type who wants to live in the countryside. She isn't a city person in her material demands and in her values like others around here. . . . But with respect to making friends and creating a social circle, well, we will have to wait and see."

Unlike their eastern counterparts, one of the most striking aspects of the establishment of private farms by westerners was the rapidity and seeming ease with which the farmers were able to gain access to amounts of land that are considered economically viable. Indeed, some of our informants were able to gain access to more land than they had originally planned, that is, more than conventional wisdom deemed feasible for a single farmer to handle. In some instances, would-be lessees were assembled by the former managers of a collective in dissolution and the contracts were signed during a single session. Those who depended on the *Treuhand* or, in the case of demesnes, the *Land,* had greater difficulties, but not as much as the *Wiedereinrichter*. This situation changed very rapidly. Those who came in mid- or late 1991 already found that most of the land had been spoken for.

The advantages that the individual farmers enjoyed over the cooperatives that had to hunt down the owners of the myriad plots that they were cultivating are obvious. This was especially true for the westerners, who were in a position to offer better conditions. The western farmers were also not fazed by the magnitude of the task and the risks of setting up large-scale farming operations, even though only the farm manager had run a farm of that size before. Rather, the westerners' difficulties were those of material and personal deprivation. First, there was the financial strain of surviving the first year without any income from the new farm. As in the case of some of the *Wiedereinrichter*, that transition was often subsidized by the income of the wife, who continued to work outside agriculture. Two of our informants were in this situation. The wife of a third informant had already come to eastern Germany before the husband and was working in a state job associated with the transition. The girlfriend of a fourth was finishing her education as a teacher while her companion continued to help in the parental farm. A fifth, a bachelor, had a long career behind him and received help from his mother. The sixth was a salaried farm manager and the seventh came to the east immediately after the last harvest in his previous farm in the west.

Given the abysmal state of the housing market in the east, with its shortages and the pitiful condition of the buildings, our informants had to make do with very simple (the word they used was "primitive") living conditions when they first came to the east. Most rented rooms or tiny apartments, and then proceeded to build new houses or rebuild old ones out of former sheep barns, office structures, and so on. One *Neueinrichter* described the family's first year in the east as follows: "It was very hard in the first year, especially that winter. We still had the old farm house on this plot. It had already been condemned as unlivable by the police. . . . Nothing worked. We cooked on two hotplates. The water pipes froze and it was terribly cold. It so happened that just in that year the temperatures sank to minus 30% centigrade. Now we are used to it."[3] In this case, a couple and their adult son had moved at the same time, leaving three daughters, including one who was still studying, behind in the west. In three other cases, the male partner came alone while the female partner and children remained in the west. The men spoke poignantly about how difficult it had been to shuttle back and forth. They all suffered from the lack of telephones and separation from their families and friends. This was perhaps most eloquently expressed by one of the farmers. "For two years, I lived a weekend marriage. At some point, it becomes too much. I was always totally stressed out, driving the car five hours there and back." Given the conditions of the highways and the demands of setting up a new large-scale farm, which could not be taken care of within a regular business day or week, commuting became unbearable. He commented on others his age whose marriages had been destroyed in this manner. In order to avoid becoming alienated from his family he would commute every week, even if he had to drive through the night. But the personal cost became too great. He concluded, "And so I finally said, 'We have to do something about this.' The children felt completely neglected and when I was at home, I was always nervous with my wife. One had to program everything into a few hours and that always led to quarrels. Under such circumstances, when one is not extremely close, then the whole relationship dies. It happens unimaginably fast. One can't leave the stress of the week behind and play the good husband. One is under constant tension and one's thoughts are elsewhere. . . . I am a man who draws his entire energy from his family. When things are working well at home, then I don't have any trouble with my work, but when I have problems, then it doesn't work. So, after a long search, we found this house [and renovated it]. . . . It has worked well. Now, when I come home at six or seven, I become a family man."

As the road conditions and the telephone network improved, the farmers were able to remain in frequent contact with their families, friends, colleagues, and business acquaintances for advice, support, and information. Family contacts in the west also gave some of our informants a sense of security. They felt that if things went badly, they could still return to the parental farm. As one informant, whose father still ran the family farm, explained, "I have a brother, but he is studying medicine and so he is not into agriculture. So, I still have an escape route. I have a fallback option that if things should go wrong here—and the risks here are not insignificant because everything is based on rent—I can always say that I still have the homestead back in the old *Bundesländer*[4] and if I can't make a go of it here, I would still be flexible enough to be able to return." Similarly the farmer from Holland spoke of not burning one's bridges. In this case, the safety net had been his wife's working in Holland so that "both wouldn't get wet jumping into the cold water." But the sense of security that the continuing ties with the west provided for the *Neueinrichter* could not eliminate the fact that they were now physically removed from their former support networks. For example, they could no longer resort to the advice of specialists, hired on a cooperative basis by groups of farmers in the west.

As the demands of the new farms increased and the prospects of success became more palpable, the western farmers relinquished both their continued involvement in economic activities back home and maintaining the option of returning to the family farm. For example, the informant quoted earlier, who initially still helped his father during peak periods, was contributing less and less of his time to the parental farm and had discussed the possibility of renting out the farm when his father retired, which he planned to do either in three or in eight years.

The westerners felt isolated and only slowly developed friendship ties with easterners. As we shall see later, they were more likely to interact with fellow migrants, with whom they felt they had more in common. The kind of socializing the men missed was drinking beer together or playing soccer. The women felt most isolated if they didn't work outside the farm and did not have children. They were particularly frustrated in their dealings with service personnel and workmen, whom they considered substandard, that is, not up to western German standards.

Only one of our consultants, the Dutchman, had been forced to move back home by 1999. The three western Germans we revisited were not only prospering economically, but had adapted well to their new surroundings. As we shall see later, they continued to expand their social roles in their communities and made fewer trips back to their previous homes in the west. Mr. Rascher, the man who in 1994 had lived alone, was now living in an attractive old house he had bought and rebuilt with his wife, who had found a job in a larger city within an easy commute. Even his parents had bought themselves a house in Bernburg and were thinking of retiring there, although, especially his mother still "felt bound to the old homeland." The other two informants also felt more at home.

The adaptation had been mutual. Thus one of our informants, who had been very annoyed at the frequent response in stores that an item she wanted did not exist, now reported that storekeepers would make note of the item and would promise to see if they could obtain it.

Access to Land

In 1994, the average size of the farms operated by our western informants, 689 hectares, was more than twice as much as the average amount cultivated by the individual eastern farmers we interviewed. This ratio holds even if one eliminates the more corporate-type farm run by the farm manager.[5] As we have stated earlier, the major paths for westerners to gain access to land was by asserting their claims to state-owned land on the basis of past ownership. Other paths were to seek land from the *Land*, from the dissolution of LPGs, from the church, and from individuals who preferred to rent to them rather than to the cooperatives or to the *Wiedereinrichter*. They enjoyed an advantage over the latter because they were able to offer higher rents. In 1991, these rents were as little as half of what they would have had to pay in the west. Those who jumped the gun and came right after the fall of the Berlin Wall were able to rent land for as little as a quarter of the going rate in the west. One farmer described them as a band that creamed off the top of the nation (*Absahner der Nation*). Being able to pay often meant that the westerners also enjoyed the advantages of relatively long lease periods: twelve to twenty-four years. While the land could still be sold out from under their noses, this was less likely if the new owners had to wait for extended periods of time before they could use the land themselves, since under German law lease agreements take precedence over sales. Even improvements made on buildings located on leased land are protected and must be recompensed after the termination of a lease.

For reasons stated in chapter 7, farmers had not purchased large amounts of land. This did not mean, however, that the new farmers were not interested in purchasing some land in addition to the homestead, for it was unbeatable as collateral for loans. The three westerners who told us that they had purchased land, purchased, respectively, 3, 18, and 72 hectares.

While it was easy for many of the westerners to secure sufficient rental land to establish a viable farm, gaining access to additional land, deemed necessary because of the uncertainty of the future price structure of agricultural produce, engaged them in stiff competition and complex and sometimes disagreeable negotiations. As late as 1994, some three years after the new farms were established, there was indeed considerable fluidity in the size of farms. As an informant with 350 hectares of land stated, "I think that a farm can never really be too large, as long as the management is alright and the farm is liquid. But I have set myself the goal of farming some 500 to 600 hectares (i.e., eight to nine times the size of his father's farm) and think that I will succeed in doing so. . . . [The large size is necessary] especially because the land is all rented, which reduces profit. So I have great interest in gaining access to more land."

The farm manager explained the kind of nerve-racking competition entailed in gaining additional land. After being able to wrap up the lease of 950 hectares of land in a single evening, gaining access to additional land required nerves and money. As he put it, "In the second farm only money spoke. Many persons (landowners) were there. We have a Dutch neighbor with whom I get along very well today. We now work together, but during that phase when we were in the process of leasing land, we were in fierce competition. We still talk about it today. He went around the village that evening putting up notices that he was paying 5 deutsche marks for each point. We had offered only four. So the next day we had to raise our offer to five as well, otherwise we would have lost the land, even though some people said that they wouldn't rent to a Dutchman. These prejudices are less evident today. Today the farmer from Holland has a large contract farm operation employing twenty persons. But the fact of having to pay the 5 deutsche marks, that raises the cost per hectare by 80 DM. If you multiply that by 1,500 hectares, that results in a huge sum."

His clash with the Dutchman not withstanding, the farm manager rented the land under close to ideal conditions. He still wondered why the former head of the collective that was dissolved at the time did not have the courage to continue the farm himself, but he felt that that fact spared him from being accused of stealing the land from a local enterprise. He and the farmer from Holland were the only bidders. He was still adding to the land when the occasion arose. For example, on the day before our interview, two persons from the west who had gained 12 hectares of land back from the *Treuhand* that had belonged to someone's grandfather came by, and offered to rent it to him. The short-term nature of the initial leases conceded by the *Treuhand* made such fluidity possible.

While in the example above, the western farmer did not enter into a head-to-head competition with a cooperative for land, that was not always the case and has occasioned considerable resentment. As was the case with the *Wiedereinrichter*, the cooperatives sometimes stonewalled the relinquishment of land or refused to engage in the usual land exchanges. One westerner told us about her annoying dealings with the neighboring cooperative. "During the first two years, we had no problems exchanging land, but more recently, since a lot of people had to be let go, they must have become angry. When we obtained new leases from owners, they gave us precisely those parcels of land and gave them to us in exactly their original state. As a result, we now have to cultivate a few smaller fields, while in the past the cooperative manager would have given the land to us in larger parcels. The committee that is now in charge seems to be unhappy with the policies of their predecessor. They have been hostile toward us. Unfortunately, there was a well in the middle of a parcel of land that we rented or even purchased, which they said they absolutely needed for their cattle. After they had first told us that we would have to take the parcels that we had actually obtained from their owners, they changed their mind and asked whether they could have the plot with the well. My husband replied that he could use the well himself if he should ever want to use it for irrigation purposes, but that they could use the water free of charge for their cattle. Thereupon, they covered the well over with dirt. My husband, who is not lazy, went with his people and a small bulldozer and cleared the well again. Now they have lost out. It is clear that some of the things that are developing here are not to their liking. But they can't stop the trend. It is understandable that the former owners now want to earn something from their land again."

As we have seen before, the situation of the demesnes in Sachsen-Anhalt constituted a special case. The *Land* not only asked for ownership and control over this land, but, wherever feasible, they planned to renew the historic structures. To this end, they forfeited rents and asked the lessees to reinvest a sum equivalent to the forfeited rent in reconstruction. Since, historically, lessees of demesnes held their rights for many generations, they had every interest in keeping up the property as though they owned it. Like many other properties, these manors have not only fallen into disrepair during GDR times, but, to the dismay of the whole population, they have been vandalized since the *Wende* by marauding gangs of youth, for many ceased to be used and stood empty as a result of the restructuring of the collectives. As the westerner who has taken over one of the demesnes described it, "A lot [on this property] was destroyed. The animals had been sold and the farmhands became unemployed from one day to the next. No one cleaned up. All the equipment and the garbage was still there. Of course, that was like an open invitation for the youngsters. They were given a free rein. Nothing remained in place; medicine for the animals, oils, and disinfectants were strewn about wildly. They forced open and rummaged in every closet. The first thing we had to do was to clear out the mess. Many of our acquaintances who paid a visit were horrified at the sight. I wouldn't have shown my feelings like that, but I understood their reactions."

Even after the short period that our informant and his family had been living in the manor, the transformation was astonishing. On the ground floor of the rather forbidding and still rundown, yet imposing stone exterior, we encountered an elegant apartment, tastefully furnished with modern cherry wood furniture, the renovation of which had cost a great deal of effort on the part of the wife to supervise. Our informant had renovated the apartments of his renters upstairs as well as a storage building for wheat. But he did not plan to take on other large empty buildings in the courtyard that were in a state of disrepair.

By 1999, the three western farmers we reinterviewed had all been able to further consolidate their position by renting and purchasing more land. The most remarkable change occurred in the case of the two farms run jointly by Mr. L., a farm manager working for a nobleman. When we visited Mr. L., we headed for his attractively refurbished historic home where he had received us before. But we were directed across the busy highway to a huge, depressingly dilapidated building complex like many that still mar the landscape of eastern Germany. We found Mr. L. in a small, cramped, Spartan office, well beneath the standards of those of his counterparts in the cooperatives or the comfortable quarters of individual farm owners.

The jarring contradiction between the condition of the enterprise's agricultural buildings and its success was soon resolved. Mr. L. told us that he had deliberately not invested in cosmetic improvements because the purchase by his boss and the family who had once owned the estate of the buildings and 600 hectares of land had not yet been finalized by the BVVG. He hoped that the assessment would be considerably lower if the assessor saw the buildings in their present state. This would be especially true if the BVVG abided by his wish that a western rather than an eastern German assessor would do the job. He and his boss, a nobleman who had once owned the estate, were still hurting from the fact that they would have to pay at all and that they would not even receive 99 hectares, that is, the maximum size of holdings that were not expropriated by the Russians in the agrarian reform of 1949. What is more, the offer of a maximum of 70 to 80 hectares of land at extremely favorable rates by the BVVG to every farm established before October 1990, regardless of origin, for which former large-scale landowners were also eligible, had been retracted because of EU objections. The EU had argued that all citizens of EU countries should be charged the same amount. In fact, the very category of former owners to which Mr. L.'s boss belonged had participated in the protest, because they wanted the return of their entire former estates.

The dispute is still in the European Court. In addition, his boss had already purchased 250 hectares from small-scale owners of the land of the former LPG that he had also taken over. "Those parcels of land," he said, "are dispersed. But with the land consolidation program (*Flurbereinigung*) we hope that that will change." It was not surprising then, that Mr. L., unlike the cooperative manager quoted in chapter 5, who only saw the disadvantage of access roads cutting up large tracts of land that would result from the program, saw *Flurbereinigung* in a positive light. There was also the possibility that he could take over the leases of the 500 hectares of land from the failed venture of the Dutch farmer if they could agree on a reasonable price. He was not the only westerner who had purchased land in the area. Neighboring cooperatives had lost more than 400 hectares of land that were snapped up "mostly by people, all of them farmers, who were displaced by urban sprawl and by the Expo 2000 in Hannover. They have received 400 deutsche marks per square meter for the land they lost and can buy a hectare of land for that sum here."[6]

Similarly, the second consultant, who runs a demesne, was able to take over 100 hectares of land of a seed-growing operation for which he had previously done the plowing. "Today, I have some 400 hectares all in all," he explained. "The firm has disappeared. Now they only test seeds here. They no longer produce them. They let the forty workers go. In the GDR each farm had to acquire the seed it sowed. Now everybody can grow their own seed [paying 6 deutsche marks per hectare for the license]. Therefore seed farms have little chance of success."

Finally, Mr. Rascher, the third western farmer we revisited, increased his acreage from 320 to 470 hectares (after losing 20 to the thruway) of which he purchased 30 and expects to get up to 100 in six to ten years. This enlargement has probably more than offset the decline in prices for crops due to decreasing subsidies and lower world market prices. Mr. Rascher believes that as long as land is still available, enlarging the farm may be a better option to offset the latter trends than intensifying production by investing in livestock raising. His father bought some too after selling some of their land in the west. His rationale for the purchases was that now "The ratio between rent and the price of land is very favorable for those who purchase land. One obtains interest rates of almost 5%." Unlike some of our eastern German informants, he was not able to purchase any land at reduced rates because he had come to the east after the cutoff date. He expressed some hope that the European Court would give them the right to purchase land at those rates as well.

The land purchases and the fact that he has only three lessors with whom to contend rather than dozens or even hundreds like many of the other farmers, including his brother back in the west, has given this consultant a sense of long-term stability. "To keep their good humor, my brother has to think up some gift to give to his lessors on their birthdays. [In the past three years], the cooperatives also have been doing things like that, feasts, Christmas cards, and small presents. There is the awareness that the lessor holds the key to success in these large farm operations. If I lose them, I lose my existence. The best management practices on the land and in sales don't help if I lose them. So one has to keep up a personal relationship with the lessor and build a moral commitment." Purchasing land, in addition to leasing, provides additional security, partially offsetting the fact that there is, at this point, no tradition of passing on demesnes from generation to generation as there is in the west. As he put it, "One can't compare the situation with the west, where the demesne renter is assured that he can keep the land and pass it on to his children. That is a different feeling."

The same informant expressed a concern about the widening gap between westerners and easterners. He thought that westerners were in a better position to purchase land than their eastern counterparts, who could not start their farms with as much of their own capital. "The relationship has become more tense than it was in 1993–94," he declared. "We have a regular gathering of a group of farmers (*Stammtisch*) and I notice that the relationship has become more tense."

Investment

Just as they were diverse in terms of their background, the *Neueinrichter* from the west were also diverse in the amount of capital they had at their disposal. But, as will already have become apparent, they all came with more capital of their own than was available to their eastern counterparts. They also had easier access to bank loans than the easterners did. Many were able to transfer their credit ratings back home to their new area of activity, and their western education made it easier to translate their plans into a language that the banks could understand. Thus one westerner had studied agriculture for five years and had worked as the administrator for a large farm in northern Germany. Even though he did not have any mortgageable assets, he had no difficulty in obtaining a loan. He explained: "That was an excellent background to be able to go to the bank and present my case. I could make clear to them what I had in mind, and as a result I also received capital. The natives here would not have been able to do that in this way. . . . If someone from here had gone [to the bank] and submitted his diploma in order to prove that he knew his trade, no one would have believed him. The eastern Germans also weren't familiar with the entire field of commerce. This experience can only be acquired through much pain and effort and at a high cost in the form of mistakes or expensive consultancies. A lot of very weak experts came who had failed to make their mark in the west and pretended that they were someone, charged high fees, but did not provide any useful information."

Much of the investments the westerners made was in buildings. For, unlike the *Wiedereinrichter*, the western farmers had no place to live. The fact that the *Treuhand* was even slower to sell land under its control than it was to rent it did not necessarily impede the initiation of construction. Just as in the case of the demesnes, the improvements made by the tenant are protected and must be recompensed if the lease ends and is not renewed. Thus one of the farmers who was allowed to rent the land that had once belonged to his grandfather was renting two of the homesteads solidly built by the latter at the turn of the century. For a total investment of some 300,000 deutsche marks, he was rebuilding an old sheep barn into a large house and the remainder into farm buildings. Another westerner was transforming old buildings located on a blind arm of the Saale River that had once housed a gardening operation into an "oasis" for his family. "The surroundings were important to us," he said. "My wife and I feel bound to nature and also like to have some peace and quiet. We wouldn't mind forsaking the use of the car, at least on weekends [to lower air pollution]. The house is 180 years old. They used to build with stone from local quarries. That was a difficult, time-consuming task, since the stones don't always fit together well. These houses are masterpieces. They have been plastered over, so that the stonework is no longer in evidence. But if one removes the plaster and looks at their history, one marvels at how they were able to accomplish something like that." In addition, he had purchased one barn and rented another temporarily which he had equipped with ventilation to store grain.

The farm manager went beyond all his colleagues in the innovative use of the surplus buildings left in the wake of the economic collapse of the eastern German economy after the *Wende*. He converted three gigantic barns and was using them to store surplus grain for a state-sponsored program under the aegis of the European Union. This had to be done under very stringent conditions and required an investment of 400,000 deutsche marks in each of the three barns. After adding the cost of a large-scale drying facility and other investments, the total sum came to 3.5 million. Although expensive and risky because the slightest infraction can lead to disqualification, the income was a welcome addition to the farming operation. He believed that he would be able to recuperate his investment within five years. Obviously, only an extremely well-connected and well-capitalized firm could have carried off such a venture. He also invested 300,000 deutsche marks of his own money into his house, a beautiful old building.

In contrast, the nobleman, a bachelor, was waiting to see how everything would turn out before he poured in large sums of money for housing and barns. Instead, he decided to rent some buildings that, like so many buildings in eastern Germany, were subject to a byzantine tangle of conflicting property claims. "[This house] was once a manor, where the Qs lived," he explained. "They cultivated what remained of the demesne for the N factory. They had operated estates in two other localities and the state had sold this one to the factory in 1942. The state now claims ownership. The factory (which was returned to the original owners) claims it too. The *Treuhand* says that it belongs to them. Since, presently, this house is under the control of the city of N, they say, 'It's ours.' They all know what they want. Yes, and the church [believes that it] will get it again and the political parties will too and so will the charitable foundation. The monastery will too. The state's policies are self-serving. That's clear. Only the private people '45 to '49 are not supposed to have a claim (that is, those whose land was expropriated between 1945 and 1949). I pay rent to the city of N. They were supposed to open a youth club here. But there are no youths in this area and so they would have had to bring in asylum seekers. They preferred to rent it to me instead. The barn where our farming equipment is housed is rented from the BGI, a real estate broker under contract of the BVVG. The BVVG was instituted by the *Treuhand*, which, in turn was created by the state. I know the shenanigans that occur around here. One can easily go into debt for a million for a new house and large barn. But we have to be able to foresee the end of the process. It could have a bitter end. And I don't need a stupid barn for self-representation."

A third farmer's homestead was a judicious mix of old and new. He tore down the old house and replaced it with a modern barn with an elaborate ventilation system. He also built a new house. But he kept the rest of the buildings around a large courtyard, merely enlarging their doors to accommodate modern farm equipment. In order to save money, he and his son did as much of the construction, including plastering and carpentry, as they were capable of handling themselves, a practice in which they had already engaged in the west. Interestingly, the man from Holland, who seemed like one of the most financially secure operators invested the least in housing and office space. To provide both, he merely fixed up a cheap, run-down, unattractive GDR administration building, built in the 1970s and 1980s in a large complex surrounding a courtyard. He said that he had given some thought about building a house in the future, but for the moment it would suffice. He was able to take over the farm buildings with most of the necessary inventory. All he had to add was a new ventilation system for the grain and a few tools.

The western farmers brought some of their equipment along with them from the west. In some cases it was hand-me-down machinery from their parents. They also bought from a western John Deere dealer who had established himself early in the region, from cooperatives with excess inventory, and from collectives in liquidation.

Some western farmers were upset at what they considered to be disinterested, sloppy, and unreliable repair service. They felt that, perhaps, the mechanics were not familiar enough yet with the high-tech electronic components of the modern western tractors. But they felt that the problem was more deep-rooted and originated to a large extent from a lack of confidence and interest in their work.

In order to take full advantage of the capacity of their equipment and to tide them over the difficult transition period, two of our western informants undertook contract work for neighboring farms and other landowners. One also cultivated 110 hectares for a neighboring seed farm, which represents a secure income. The second, the Dutchman, operated the largest contract labor firm in the area, using his pool of capital-intensive, special machinery to undertake tasks for which his neighbors were not equipped. He even did all or most of the work for those who had not been able to amass enough land for a viable farm and therefore engaged in farming only on a part-time basis, perhaps expecting a son or a grandson to eventually take over the operation. He also had a contract to cultivate the land for his Dutch partner, who ran a dairy farm. Finally, in order to further diversify, he also spread liquid manure in the fields for his clients. For this purpose he was using western tractors and combines, for, even though he thought that the quality of the Soviet bloc machines was sometimes quite good, he had only been able to use one of the tractors he inherited. He figured that a combination of working for others and working for himself represented a good mix. In some years, contract farming could be very frustrating when wet weather conditions forced him to return to the same fields several times before the work could be completed. He thought that during a dry year, his own harvest might suffer, but that farming for others was time-efficient.

Like the *Wiedereinrichter*, the westerners sometimes hired others, particularly the cooperatives with special equipment, to undertake specific tasks for them. They also belonged to machine pools but made little use of them. The frequent pooling of machines practiced in the west, of the kind we already encountered in the example of one easterner, was not practiced in the new setting. The farmers claimed that they needed their equipment to cultivate their own large farms or, on occasion, to work for others on a contract basis.

Personnel and Division of Labor

In a manner similar to the *Wiedereinrichter*, the independent western farmers were all personally involved in actual farming. This was true regardless of the size of the farm. In contrast, fewer of their wives were, or expect to be, involved in agriculture. The reasons are varied. In at least two, if not three, of these cases a more traditional western German division of labor, where the employment of women was lower than in the GDR, may also have played a role.

In the two cases where the wives were personally engaged in the farm, a western German and the Dutch woman, the decision entailed a drastic career change. However, neither wife engaged in physical labor on the farm. Other kinsmen, usually fathers or brothers, also contributed their efforts to the farm.[7]

Due to the larger than average size of their farms, the western farmers tended to hire more full-time, ongoing workers than the *Wiedereinrichter*, but the number of persons working per hectare of land was far lower than would have been the case in the west.[8] When the need arose, the western farmers also hired unemployed or retired workers from the locality as seasonal workers. The farmer from Holland, who worked with nine workers, all formerly employed by the LPG, told us that, in contrast to another of his Dutch colleagues, he had hired local personnel rather than bring people from Holland, because he didn't want to create "a Dutch colony. It is also not necessary, one can hire good people here. It doesn't make sense to bring people from another country when the people here are unemployed." He expressed surprise, however, that the eastern Germans refused to engage in certain types of work such as harvesting hops. In desperation, one of his clients resorted to the seasonal employment of twenty Polish workers to undertake this task.

In general, our informants were very positive about their workers, although some claimed a lack of initiative and self-confidence. They were also used to a different rhythm of work. The Dutch farmer commented that he preferred older workers who remembered work in precommunist days. They would complete a task no matter how much it lengthened the workday. He was also very pleased with the early retirees, who got a kick out of working with all the new machinery. Interestingly, he felt that his equipment operators were more inflexible. Unlike their counterparts in Holland, they were more reluctant to attempt to make minor repairs of their equipment since, in the past, they had relied on members of the LPGs' specialized mechanic brigade to undertake such tasks. Although he had hired a mechanic too, that mechanic also worked as a field hand, and so everybody needed to be flexible and able to do whatever came up. The tractor drivers were learning fast.

One informant also commented about differences in the degree of loyalty to the enterprise. One of his workers had left as soon as someone offered him a higher wage. He felt that it would not have been so easy to hire away a western farmhand. Other western farmers expressed no such reservation. One informant praised the "exceptional dedication to work and diligence" of his employee and is happy with the way in which he identified with the farm and worked on his own. All he had to do was give him the tasks and he did them without any kind of supervision. Similarly, another informant commented that one of his workers, who in the beginning seemed to despair of ever being able to master the handling of a large new tractor, was now confident and often suggested new ways of accomplishing a task. In general, then, finding qualified workers and training them to undertake new tasks was not difficult.

In 1999, at least two of our western informants were, like their eastern counterparts, hiring more outside help. Thus the demesne farmer kept on the apprentice he had hired even after he had stopped raising pigs because "The apprentice reduces the work load of the farm manager. It is a question of the quality of life. The number of workers depends on the profit we make. The farm size we have could also be cultivated with only one worker." He feels less pressured than in 1994, because he is more secure in his knowledge of the profit margin. Similarly, Mr. Rascher hired an apprentice in addition to the worker he had already hired because "This is a social obligation of sorts. I am not sure whether it is an optimal solution in terms of cost as the officials claim it is. We could also have hired *Arbeitsbeschaffungsmaßnahmen* (ABM) workers, that would have been cheaper." Since the training program stipulates that the apprentice must have a broad-based education, Mr. Rascher has made an arrangement with another farm to introduce the apprentice to cattle raising. "It would be better if an apprentice would switch to another farm every year (of the three-year apprenticeship) but here it is not done, because there are not that many farms that take on apprentices. It is also a matter of cost." Mr. Rascher receives a subsidy of 3,000 deutsche marks a year from the state, amounting to about a fifth of the 15,000 to 17,000 deutsche marks the apprentice receives per year. At 16.10 to 17.50 an hour before taxes and benefits, the wages paid by the westerners to their permanent workers were somewhat higher than those paid by their eastern counterparts and by the cooperatives.

Crop Production and Marketing

The western farmers in our sample, all of whom came from regions with more rain, had to adjust to a drier climate. They were surprised that crop yields were some 30% lower than in their regions of origin, requiring larger amounts of land. They had to employ techniques such as plowing in the fall and leveling the soil in the spring as well as reducing intercropping (which takes moisture out of the soil) that improve water retention in the soil. Some were also planning to establish irrigation systems. One farmer also commented on the high degree of pollution of the soil and the water (he did not say from what), which could prevent them from meeting EU norms in the production of certain herbs. At the same time, the westerners were pleased to note that much of the soil was less heavy and therefore required less labor and reduced equipment maintenance.

Like the other farmers, the westerners have taken advantage of the region's excellent soils while being cognizant of the limitations imposed by the dry weather conditions. Strapped for cash because of the high start-up costs, most of them have tried to minimize risk and employ strategies that equalize labor demands over the growing season and minimize extra labor costs. As one western farmer explained, "The grain crops ripen one after another. Peas [for fodder] are sown in spring, then one has time in the fall for the grains. At harvest time, peas compete somewhat with wheat, but it is still manageable. Winter barley ripens first, then rape seed, then rye, followed by wheat. At the end of the wheat harvest come the peas. This schedule also evens out the risk. The sugar beets come at the end. We have them harvested by a contractor, which makes economic sense because the harvesting equipment is very expensive and we are terribly overworked at that time because we are beginning the next round of sowing. Because of the high number of contractors, there is considerable competitive pressure, which has a beneficial effect on cost for us." Because this mix is optimal in so many respects, he did not foresee specializing in any particular crop.

As we have seen earlier, the dry climate, while reducing yield, also results in rapid maturation, which enhances grain quality. The same informant explained: "This is an important factor for us. . . . [The quality] is comparable with Canadian and American wheat, which used to be imported [in the west] in order to improve local grain supplies. The flour mills mix our wheat with the inferior wheat. The demand comes from the Benelux states and England, so we have can market our grain under favorable conditions." Well-established traditions of growing particular crops in the area and well-established markets for these crops certainly constituted an important factor in determining what to grow. Indeed, some of our informants were speculating that the association of particular crops with certain regions would be further consolidated in the future and production quotas shifted from regions that are less suited for a specific crop to those that are best suited. Thus, one informant was reluctant to sow rape in fields where he might be able to grow sugar beets in the future, should he be able to increase his quota because rape reseeds itself and becomes a weed that interferes with beet growing.

Although all the westerners reserved small areas for experimental crops, at this point, only the manager of the very large estate felt secure enough to be able to engage in such experimentation on a larger scale.

The opportunities presented by the introduction of sunflowers is instructive. While one farmer saw growing sunflowers as a risky proposition because he had been warned that the area was marginal for the crop and that a wet fall could lead to a total loss, mitigated only by crop insurance, the farm manager felt that the narrow strip of land in the rain shadow of the Harz Mountains was "predestined for sunflowers. At the moment it is a business that is unbelievably good. With all these General Agreement on Tariffs and Trade (GATT) negotiations with the European Union and things like that one thinks that everything is going down the drain. But the sunflower is definitely on the rise. Because of the soy bean harvest in Brazil, where 10 million tons less were produced than projected and the floods in your country the entire production has been lowered. As a result, we have a contract for the entire production for the coming year, something unheard of in western Germany. Last year, it began with rape seed and this year with sunflowers, which we sold for 52 deutsche marks per 100 kilograms delivered at the farm [17 more than last year]. But then, suddenly, things can go downhill again; one can't understand the entire politics." When we asked him whether that meant that one had to be much more flexible in one's productive strategies than in the past, he agreed. "Exactly. Before, when I was an apprentice, it was a simple matter. Sometime in winter an agricultural consultant would appear on the farm and a crop sequence would be established. In the west we really had only one crop sequence. We already knew twenty years in advance what we would be growing on any particular plot. It is my opinion, that, today, we can only produce what is currently in demand. At this point, we can earn our money with sunflowers and so we have to go all out." The manager did not just react to world demand, but also planned proactively. Thus in the future he hoped to take advantage of a pond with three springs to irrigate a special 6-hectare field for potatoes. Since the price of potatoes had plummeted, the manager's plan was clearly based on longer-term projections, rather than on prospects for immediate profit. Such a strategy also underlay the plan of the farmer from Holland to grow quality potatoes. "My father always said, 'When you begin something, then you have to begin when the price is the lowest.' I think that because the price for potatoes is so low now, many will stop growing them. So I think and I hope that if I start now, then the price will go up, if not this year then in the future. I think that there are enough potatoes but not of the quality that people expect. That is what we are trying to do now."

The farm manager's orientation toward a world market was echoed by the nobleman who had spent many years in southern Africa and had a brother in South America. He was toying with the idea of planting medicinal herbs and establishing a textile dye producing plant with low allergenic properties for persons with sensitive skin. His brother's factory would provide a ready outlet. His plans were in consonance with another informant's musings about the possibilities of production for industrial use. Indeed, such an orientation would allow the farmers to overcome the obstacle of EU overproduction that underlies its policy of leaving a large proportion of land fallow. Production on such land would be permitted to the extent that it not increase existing surpluses. France, for example, dropped its tax on fuel produced from rape seed to foster its production, thereby capturing GATT-established quotas for agriculturally produced fuels.

While the same sources of advice about what to plant and what inputs to use were available to the western farmers as to eastern farmers, an important institution in at least some regions in the west, consultancy through membership in a consultancy association, had not (yet) been introduced in Sachsen-Anhalt. One informant explained: "In Lower Saxony some one hundred farms would form a consultancy association. The association would hire a consultant who had a degree in consultancy entailing a three-year study period to obtain the degree of consultant on agricultural issues. Each farm paid a flat rate and an additional amount per hectare of land they owned. Fifty percent of the expenses were shouldered by the chamber of agriculture, a semigovernmental agency, so consultancy was cheap for the association members. The farmers themselves determined what problems were to be prioritized by the chairman of the association. They would engage in experiments themselves and were in close contact with the [semi-] governmental agency. The latter and the association worked closely together. In turn, the consultant hired by the association was able to gain insight into all the farms in the association and thus, as an independent agent, he was well informed about their problems. Here that kind of thing has not been developed. People said that consultancies should take on a more private form. As a result, if I hire a consultant, it costs me some 120 marks per hour plus traveling expenses. That is the usual fee. That means that I have to be able to formulate my consultancy needs very concretely, and if I can do that, then I really already know the answer.

"Often, all we are trying to do is to reassure ourselves [that we are on the right track] and all sorts of positive things develop out of such conversations. . . . But for that, I wouldn't normally hire a private consultant. I would really be quite confident about my judgment. Frequently the issues are very concrete technical ones, such as whether to use one product or another. A product that led to good results under specific conditions during a previous year may not have the same results during the present year because the conditions are different. The association consultant would usually have encountered the same problem a few days earlier with another member of the association. That would not happen here [because a consultant would work in a very much larger area]. Only since I have begun to work here have I become aware [of the importance of this type of consultancy]. My siblings, who have considerable expenses because they are required to pay very considerable contributions to the agricultural chamber, tell me how lucky I am that I don't have them, but they don't realize the advantages that they have." In this case, as in many other respects, the tendency was toward more individualistic capitalist solutions while cooperative or collective arrangements were receding.

By 1999, the three western farmers we reinterviewed had made relatively few changes in the kinds of crops they produced. The nobleman had given up the idea of planting dye-producing plants. Mr. L., the farm manager, had further developed growing potatoes for a supplier to McDonald's, initiated in 1994, into a major component of his production and was presently cultivating 120 hectares with these potatoes. The venture was not for the faint-hearted. At present, only three other farms, all cooperatives, dared to grow these potatoes and each had planted only 30 hectares of them. It entailed growing a special variety of potatoes under exacting conditions. Another informant had told us that one farmer had lost the entire harvest to a disease and had not even been able to sell them as cattle fodder. But Mr. L. was unfazed. He said, "We produce for a Dutch firm that was acquired by Americans. From here, the potatoes are shipped to Holland, where they are converted into french fries and then they come back or are sent to Poland or wherever. They have pure white flesh and are very long. They need a lot of water and we have a dry climate here. But we have a lake here and in Z. we have wells, so we can irrigate them. The variety is very sensitive and one needs a green thumb to grow them. The plants are susceptible to rot (*Krautfäule*) that is promoted by high temperatures and the humidity from irrigation. The potatoes are then not storable. And then there are all the animal pests, for example, the wire worm (*Drahtwurm*) that eats its way through the potato, especially in July when they ripen. One has to be just on time with pesticides. We spray them every Monday. The deals are stringent. When the quality is not up to par, there are painful deductions. When the potatoes are longer than 10 centimeters, we receive a nice premium. Also, there are limitations to how much can be planted. The bottlenecks are the irrigation machines. Also, one can plant a field in these potatoes only once every five years. But in the past six years it has worked very well."

Mr. L. had initiated the venture against the advice of his agricultural advisor. "I am still one of those persons who decides to do something because I have a gut feeling about it (*aus dem Bauch raus*)," he declared. "When I do something, I do it because I am convinced it is the right thing to do. And sometimes our agricultural consulting service says, 'Let me calculate whether this is really that important.' It's good if someone puts on the brakes. But when one's profession and one's hobby is one and the same thing, as it is in my case, then it is difficult just to write down the right numbers and not to decide from a gut feeling. It was like that with those potatoes. When I got to know those people I said to myself, 'I believe that the way these people do business is 100% correct.' And it works. So I said 'Let's do it.' I didn't stop to make all sorts of calculations. I did it out of a gut feeling. If we are shipwrecked by it, so be it, we won't be poorer, but will be enriched by the experience. As it turned out, it brought us a lot of money." He was less lucky with an experiment growing carrots for a large-scale distributor. He put 15 hectares of land at the disposal of a group of investors who paid for his labor and for the land rent, while he himself planted an additional 15, assuming the risk himself. The venture failed, but he still broke even. "It didn't work out because enormous storage facilities for vegetables had been constructed in the middle of nowhere with financial support from the EU. The people who managed the venture were, in my opinion, incapable. A number of eastern firms were involved. So, suddenly the place was full and there were still carrots out in the fields and nothing had been sold. November 10 arrived and it became cold but on November 15 our carrots were still in the ground. So we tried to sell the carrots elsewhere. I ended up visiting all the horse stables in the area and dumping the carrots there just in order to get rid of them. The last 15 hectares we didn't even bother to harvest. We had to plow them under. We neither earned nor lost money. It was the wrong decision to make, but the partners weren't the right ones either."

A final venture did not involve crops but windmills. As he admitted, "As a farmer, my hackles are raised, they look awful, but financially it constitutes another source of income." Mr. L. was not content with leasing land to an outside venture like the cooperative mentioned earlier, but has participated in the venture himself, constructing four windmills, an investment that should pay off within seven to ten years when the machines are amortized.

Mr. L.'s entrepreneurship is admired by his peers but their managerial strategies differ. According to another western farmer, "Those who did not experiment around and have only planted wheat have earned the most in the past five years. In other words it is best to depend on the state [rather than plant crops that are not subsidized]." When we said that in the long run those who had experimented would have more flexibility because they would know how things could be done differently, he answered, "That is precisely the point, to accumulate managerial expertise. One learns from one's mistakes. I hope that I don't repeat those I made in the beginning. But one is not forced to make mistakes if one only cultivates subsidized crops, wheat and things like that. One only needs the means of production and the sale of the product is very simple because it is very homogeneous and controllable. In contrast, quality control is very difficult with potatoes. They become discolored. It is a different product, less storable, less homogeneous." Finally, the third farmer, Mr. Rascher, has decided to join a commercialization cooperative and plant potatoes to supply the starch industry.

In conclusion, the western farmers in the east appear to be somewhat more attuned to the world market. They are more aware of specific opportunities and limitations and have the connections and, in some instances, the capital, to take advantage of the opportunities. Some also had unique connections both in western Germany and elsewhere that enabled them to explore alternatives. Tradition nevertheless also plays a significant role because of environmental constraints, the association of the region with particular crops in the minds of middlemen, and the agricultural politics of Germany and the European Union. Within these constraints, individual farmers try to carve out their particular crop mix. It would be easy to conclude that cultivation is only market-driven (modified through EU intervention) if one didn't know the background and knowledge of each farmer, which informs the way in which he analyzes the current situation as well as his outlook for the future.

Livestock Production and Marketing

As a group, the westerners were less likely to engage in livestock raising in 1994 than their eastern counterparts. In fact, only one raised animals at all, although two had more than vague plans to do so in the future.[9] Interestingly, the farmer who did engage in animal raising, the farm manager, made the most surprising choice, namely, sheep raising, which had practically ceased in the area. The choice combined the more corporate direction of some of the largest farms set up by westerners and the family tradition of the *Junker* owner. He raised lambs under contract with restaurants in the west, butchering them in the enterprise's slaughterhouse in northwestern Germany and at the same time continuing a breeding tradition initiated by one of the owners' great-grandfather. Specimens of this herd were well known as far as Scotland, South Africa, and New Zealand. Our informant took the sheep along with him to the east, since his successor in the west had no experience with sheep raising and entrusted them to a local shepherd and his wife. Ironically, the enterprise was doing business in the same way as the LPGs had before the *Wende*. For then, too, local consumption of mutton was minimal and most sheep, heavily subsidized by the state, were exported to the west. A further irony was that the family of another of our western informants was forced to *abandon* sheep farming on their farm in the west as a result of the glut on the market that resulted from the liquidation of eastern German herds, a fact that contributed to his decision to move to the east. He had no immediate plans to start raising sheep again in the east.

The farm manager had also attempted to reestablish pig raising in a barn that had just been completed before the *Wende*, but had been empty ever the since the precipitous sale of all the livestock by the manager of the erstwhile state farm. However, the town had vigorously opposed the move because of concern about the smell and noise, and as he put it, "In Germany if one is located within a settlement it is almost impossible to start raising animals again once one has stopped, if the right people pull some strings." The farm manager felt that pig raising would eventually revive from its present slump. He felt unjustly beleaguered, for had he not provided work for ten local persons and could hire two more if he were to be allowed to raise pigs? In contrast, the state farm had only raised the young cows in the locality, for it had other operations elsewhere, and the facilities would not have been adequate for a full-fledged dairy farm, which he regarded as the most lucrative branch of cattle raising. In addition, there was no milk quota to take over with the farm and the investment would have been prohibitive. Another farmer had included pig raising in the proposal submitted to the *Land*, but was reluctant to proceed because of the low price of pork.

In 1999, the farm manager was still raising sheep but more as a hobby than because it was a money-making proposition. He no longer delivers lambs to hotels directly. "One of the hotel managers I used to supply tells me that he only buys on the Internet now," Mr. L. commented sadly. "And in the east they look more at the price than quality. In the west it's different. When the animals are ready to be butchered I call a livestock dealer. I don't know where he sells them. The last ones went to a slaughterhouse nearby." Meanwhile, the demesne farmer, whose initial plans submitted to the *Land* included raising pigs, was running into difficulties. He had already engaged an apprentice and had set up a barn for 400 pigs, but things did not turn out as planned. In addition to the drop in prices, he had made mistakes in barn design and in marketing. He explained: "I had visited several individuals who had this type of installation and they said that it was working well. But two years later, I read in the newspaper that a number of people had failed using this design." He was now toying with the idea of building a small biogas plant to generate electricity from the pig dung. He figured that, in theory, he should be able to earn an extra 15,000 to 20,000 deutsche marks a year in this manner, since the SPD/Green coalition in power fostered alternative means of producing energy, but he was wary of repeating earlier mistakes.

Suppliers and Markets

Immediately after the *Wende*, the conditions for the commercialization of eastern agricultural goods did not inspire confidence in the future of fair commerce. Although they themselves were not the direct victims of abusive marketing schemes, and, at worst, suffered losses in the west from depressed prices, the westerners nevertheless expressed dismay at the rampant fraud of that early period when entire animal herds seemed to vanish in thin air, orchards were cut down in their prime, and shady "advisors" roamed the land. The loss of markets to the east, the decrease in local demand due to the cornering of the retail market by western chains that tended to privilege western suppliers, and the virtual demise of the eastern German food-processing industry also did not inspire confidence. In spite of all these ominous signs, however, the markets for certain agricultural goods turned out to be amazingly favorable to the farmers in the east both in terms of the price of inputs and in terms of what they could obtain for their production. One informant described his astonishment: "I did not expect the good conditions for purchasing inputs. The consumer cooperative (*Raiffeisengenossenschaft*) became firmly positioned in the market. It purchased [facilities] from the *Treuhand* on a large scale, possibly, in part, through fronts. I imagine that they must have been told that for every two or three plum branches they would have to take over ten inferior ones. [Even the miller that came from the west, took Raiffeisen as a partner]. But Raiffeisen did not realize that there would be so many traveling salesmen (*Streckengeschäfte*) who come around without even having to open a local office. The representatives can come alone because the farms are so large that they can service a larger area in a day than they could service in the west in an entire week. They can thus work on a smaller margin and so we can profit when we buy fertilizer or pesticides. Unexpectedly, we also command higher prices on sales than in the old *Bundesländer*. This is true in spite of the fact that the relationship between consumption and production should have been less advantageous for us, and therefore the prices should have been lower." His wife, a former official in a government agency, continued: "Factors have combined in unexpected ways. I would never have thought it, but because of the fact that during the GDR the grain-purchasing operations were so enormous, they also had gigantic silo complexes. [Those who own or manage them now] must try to fill them and so they outbid one another to accomplish this, because it would not make economic sense to leave them empty. And so the farmers can obtain good prices."

Not all the farmers could take equal advantage of this situation. As the same farmer commented, "I naively assumed that no great difference would be made between a 290- and a 1,000-hectare farm. They were, after all, all large farms. But I found out that major differences are made [in the prices for inputs]. The small guy can only guess and sometimes hears about it through roundabout ways how well the large guy is able to make his purchases."

The availability of inputs at reasonable prices was further enhanced by the fact that the farmers in the east now have unlimited access to the products of the eastern German chemical factories that have survived the restructuring after the *Wende*. As the farm manager explained, "Let me give you an example. We work exclusively with liquid nitrogen fertilizer and here in Piesteritz, some 40 to 50 kilometers away, is the largest currently producing factory in eastern Europe. That factory already existed during GDR times, but the GDR never worked with liquid fertilizer, it all came to the west. I still remember that during western times (i.e., before the *Wende*) the trucks would come from Piesteritz. Those were dyed-in-the wool truckers. They had to be in possession of a party card. Well, the SKW Trosberg, a Swedish-German industrial corporation, has bought the plant in Piesteritz and has transformed it into a factory with the latest equipment. It is fascinating to compare how it looked before and how it looks now. It is properly laid out with special roads for loading, all computer-controlled, not a drop is lost. Before, the fertilizer was running all over the place, one couldn't bear to look at it, it was so dreadful. Now everything is in perfect order, and I obtain liquid fertilizer for 17 marks per 100 kilograms, while in the west we pay 18.20 for our farm size—we have 500 hectares there. So that in the last analysis the difference [in profit per hectare between production back home and here in the east] is narrowed considerably. Even though I have much lower crop yields here [than in the west] I am better off here because of all the advantages of running a larger operation."

The potential for selling one's agricultural output to local processing plants has also improved, but only for certain products, while for others privatization has meant mostly the elimination of previous capacities. In 1991, the farm manager transported all his production to the west "because the prices had totally crashed. We had two of our own trucks transport everything to the west because the prices were higher there. By the 1992 harvest, the agricultural dealerships finally realized the opportunities around here. They established themselves here and created the infrastructure to store grain. For example, they rented large barns from the LPGs so that they could store the grain prior to shipment and thereby get into the business. There was no dramatic change during that harvest, but afterwards when they offered the grain they had purchased on the market, they realized, 'Man, this quality offered over there is superb.'" While during that year, he sold to an intermediary, he had now established direct ties with a client who bought his grain from that wholesaler and was obtaining an excellent price, further narrowing the difference in the profit margins between the east and the west. Now he also had the option of selling his grain to the privatized, large, local western-owned flour mill where, thanks to the large quantities he could offer, he also received a preferential price.

FIGURE 8.1. The regional sugar factory near Bernburg.

In the case of grain, the farmers also had other choices. But privatization of industries associated with agriculture did not always have such a salutary effect. One of the approaches toward industrial restructuring in eastern Germany which was adopted by the German government is what Gernod Grabher (1992) and others have called the "Castle in the Desert" approach. In this case, "model" western firms were established at great cost in terms of state development funds in order to provide examples for local potential entrepreneurs to emulate. The results of this approach, which was often carried out without taking into account more locally based alternatives, were often disastrous. An example of such an approach to eastern German development (by no means the worst) is the gigantic sugar refinery established near Bernburg that started production in 1993. As we have mentioned in chapter 6, although it is one of the largest and most modern ones in Europe, if not the world, it was incapable of handling the bumper sugar beet harvest of 1993. In contrast with the flour mill, which retained the old processing facilities while planning and carrying out the construction of a brand-new, higher capacity facility, all the older and smaller sugar factories were closed. At the same time, antidumping laws and high transportation costs made it impossible to find buyers elsewhere in Europe. The farm manager, who, as we have already seen, otherwise had a very positive view toward western industry, was devastating in his critique. "[That case] is an example of how things can go wrong when one concentrates everything in one spot and builds on such a large scale. It is easy to conceive something from scratch in an empty space ("auf der grünen Wiese"), but it is much more difficult to convert these ideas into practice. It is inconceivable for us that this enterprise would not be capable of dealing appropriately with the situation on the ground. All that would have been necessary would have been to continue production in one or two of the old factories. In that way they would have had a good start. That might have cost them a million more, but now their image is tarnished; they have all of us [farmers] on their case, and they must also pay us an indemnity for the sugar beets that are rotting in the field. This strategy is also being employed in [another location] where a similar enormous complex is being built. Of course, the result is that four or five factories will be closed in the west. You can imagine what effect that will have on the small farmers there. In the beginning, the people in the east damned the Wessis and now, from the vantage point of agriculture, the tables have turned." This form of privatization without concern for the repercussions in each locality was replicating the problems associated with central planning and gigantism characteristic of the former GDR regime.

Many industrial outlets for agricultural products that were closed after the *Wende* have not been replaced at all. The canning factories in the region were all closed with the exception of a small factory that produces jam and one that cans pickles and fruits, which by 2000 was struggling to survive. It was indeed difficult to find any processed food of eastern German origin in the supermarkets. Some of our informants spoke about the possibility of a factory that would produce and package ready-made salads. But because of the problems of ownership and regulatory restrictions, it might take up to two years to obtain the go-ahead.

Many of the more specialized markets could only be found in the west. The farm manager, for example, who, as already noted, continued to sell even wheat directly in the west, made regular trips there to market his sheep. He remarked, "The rams that we raise—there are always some twenty of them—we of course continue to market in the west. We travel there for the auction in Paderborn near Kassel, one of the most famous in Europe. Buyers come from outside Europe as well, and I will continue to sell them there because here the demand is less for top breeding animals than for animals for fattening."

One of the ways in which farmers could gain more control over access to inputs and commercialization of outputs would be through producer cooperatives or shareholding companies where the farmers hold a controlling proportion of the shares. There were, indeed, examples where such an approach had positive results. Thus, one of our informants contrasted the local sugar refinery where the corporation was the majority owner, with one in an area nearby. In that plant, the farmers had considerable say over the management of the factory because they hold a controlling interest. For example, such a situation would permit farmers to sell sugar beet production quotas to one another and commit the factory to process sugar beets above the contracted amounts even if at a lower price. The fact that the local factory broke its promise to process the entire production, regardless of the quantity, was unique in this informant's experience. His father had farmed for forty years in the west and had always been able to sell his entire production.

In our area of research, cooperative arrangements among independent farmers have had mixed results. One such cooperative was founded mainly to commercialize grain. Although the grain-marketing cooperative had a membership of eighty that include *Wiedereinrichter* and cooperatives, fifteen members were specially active in engaging in joint purchases and sales. The rest participated in occasional transactions with dealers as they saw fit. The organizer described the rationale as follows: "The whole idea is of course to gain power in the market. Particular [transactions] must be done in large lots. When we make a deal with a mill in England, we can't just offer a thousand tons of wheat. They would tell us that they can purchase bird feed at the grocery store. Things change when we can offer five or six river barges full of grain. Then they become interested and it becomes worthwhile. Such transactions are not simple. One has to take all sorts of legalities into account. We achieve market power and engage in such business deals together with a dealer who initiates them and carries them through. He also shoulders the responsibility. However, it is important to ensure a high quality."

According to one of the smaller farmers in our sample, such cooperatives (*Erzeugergemeinschaften*) exhibit severe limitations. First, they tend to be spearheaded by the *processing industries* rather than by the farmers. The industries' interest in doing so derives from a law regulating commerce that concedes subsidies to such cooperatives for building silos and loading docks. While, in theory, the subsidies are available to all members of a cooperative, in practice, only the processing plants are in a position to take advantage of them because the other members lack the courage and/or the capital. "They therefore initiate the creation of a producer cooperative themselves even though such a cooperative should be oriented against them." One of the organizers of such a cooperative disagreed with this assessment. He claimed that, although in the cooperative that he helped organize the mill had the original idea, the farmers created an independent organization that cooperated with the flour mill but also sought outlets outside the region, which, in addition to quality wheat, also included buyers of quality rye and brewing barley. He also claimed that the cooperative, which had a membership of eighty in 1993–94, had been able to gain access to markets that would have remained closed to the individual producer.

Besides resembling a 'company union,' such cooperatives also appeared to suffer from a lack of solidarity among the farmer members who, when in a position of authority within the cooperative, took advantage of opportunities that came to their attention because of their position for their personal gain rather than making the benefits generally available to the entire membership. The first informant concluded sadly, "I don't know whether this [type of behavior] is typically German. The tradition of solidarity (*Solidaritätsdenken*) is not very well developed and everyone thinks that they have to take matters into their own hands." Finally, the two informants concurred in the perception that the dealers did their utmost to subvert joint action on the part of their clients. This was accomplished by denying better rates to marketing cooperatives even when they offered the same quantities presented by the large farms, or by paying high prices to individual small farmers who made separate deals with them.

Community Relations

The self-conception of the westerners, including their perception of their historical role, the way westerners and easterners view one another, the way fellow western migrants view one another and their countrymen who remained in the west, and their perceptions of their respective economic and social systems define strategies, shape their social networks, and inform relations in the community and the wider region.

In general, the perception of the westerners appeared to be that stereotyping of them as "Other" was no longer as prevalent as the media made it out to be. Nevertheless, it also was clear that westerners continued to have some difficulties establishing more than superficial interpersonal ties with easterners. A major factor was certainly the short period since their arrival and the lack of free time for socializing during this period of struggle to establish themselves. In one instance, the fact that the informant's fiancée was still back in the west also played a major role. So the deeper relationships were often still with family and friends back in the west. Our informants often made special efforts to go back home every few weekends to maintain these relationships. The relative geographic propinquity (a three- to four-hour drive even under less than optimal road conditions) made this possible. The new ties established in the east were also often with other westerners. The voluntary associations promoted by the westerners tended to be dominated by them, and more casual visiting or pub commensality also tended to be segregated. As the farm manager explained, "Unfortunately, for the moment, the good contacts that we have established are with other Wessis. To be blunt, we have Australian conditions here. We will drive 40 or 50 kilometers to visit the next [Wessi-owned] farm, even though the nearest farm—unlike in Australia—is only 2 kilometers away. We bypass many a farm in order to visit someone with whom we have something in common." Another westerner elaborated: "It is strange, but the fact remains that we have interests similar to those of other persons from the old *Bundesländer*, even in matters concerning agriculture. The people here in the new *Bundesländer* do, after all, have different problems." This informant played squash in a distant town with three other individuals, one farmer and two businessmen, all westerners. This preference for western friends was true, in spite of the fact that westerners often perceived easterners as being more caring and welcoming than might have been the case had the westerners established themselves in a strange community in the west. At least one western informant ascribed the lack of contact between westerners and easterners to the arrogance of westerners. He said, "The *Neueinrichter* often have a very high opinion of themselves and are outraged by what they experience here. Even though it is one Germany, there has been no mutual accommodation, and they cannot empathize with the problems of the local people."

When the westerners did attempt to engage in a dialogue with their eastern counterparts, they were sometimes disturbed by the fact that the latter were not fully appreciative of the merits of the western political and economic system. They had not necessarily experienced an improvement in their material circumstances and felt that the present government misled the populace as much as those in the past. The westerners were shocked and hurt when easterners told them that they wouldn't mind if the Berlin Wall would be erected again. They also were uncomfortable with the strong association some easterners still seemed to make between festive occasions and politics. Thus one western informant thought it inappropriate to invoke political symbols, such as the national anthem, during a harvest feast. This same woman perceived that a criticism of the GDR was taken by some easterners as the denial of their own worth and their contribution to the world.

As we have seen earlier, part of the appeal of coming to the east was the prospect of recapturing the roles that their ancestors had once played in the east. Some of them had lost these roles entirely when their family estates were expropriated, others had continued their elite role as landed aristocrats, but on a smaller scale and in an area where the tradition of landed gentry was less well developed. As we have already seen, their dream of gaining back what their families had lost was to a large extent thwarted, but they were given priority in gaining access to whatever land was available. Farmers from bourgeois backgrounds were also attracted by the prospect of achieving a more exalted social position within the community and the region. As indicated above, one took over a demesne, another chose to refurbish a historic house in an idyllic setting, and a third chose a location that would enable him to hunt.

Both farmers from noble families, or representatives of such families, and the others viewed themselves as standard-bearers for an all-German tradition of promoting civil society, sports, music, community festivities, the church, and local and regional voluntary associations[10] and regular informal gatherings (*Stammtische*) at restaurants or pubs. The renter of a demesne and the two westerners who came from families who once owned land in the area regarded themselves as successors to and restorers of historical traditions associated with specific estates and/or family lines. The renter of the demesne was well aware of the fact that before World War II the demesne lessee was not only the major employer in the community but also the patron of the church. As he put it, "He was a functionary and hence carried responsibility. In the localities in the west

FIGURE 8.2. Women's choir in a small town near Bernburg.

where enterprises with this legal form were continued, there has been a seamless transition, and to this day, the [demesne lessees] have continued to fulfill many of these traditional obligations. To the extent possible, I also intend to follow this goal and I have noticed, especially among older people, that they expect as much. They say, 'Now the demesne is back again.' It has already begun with the church. Its surroundings were unkempt and we were asked whether we would be willing to lend our truck and a worker to clean it up. I welcomed the opportunity because, as a newcomer, I would like to become a part of the community. I believe that [in the community nearby] the aristocrat, or rather [his manager] will try to do the same thing."

When we interviewed this manager, he concurred. He had felt obligated to contribute to the restoration of the church in order to reestablish civil society in the village. He elaborated: "In western Germany this [role] is more common. One has grown up in a community and therefore knows the people by having attended the same schools or the same church or clubs. Here the associational activities were pretty much destroyed after the *Wende*. The people from the east say that friendships and easy contact, having beer together, was destroyed with the *Wende*, and that jealousy has taken its place of those who want to establish an enterprise. Not only did [the entrepreneurs] have to work harder [and therefore have less time] but easy commensality has also disappeared. I should say that associations have not died completely in this locality. The soccer club is a positive example. It is well organized and functions well. Both of our sons play there and I attend the games when I have the time. Ten to twelve fathers are usually there and we get to talk. In that respect there are no problems. I find that that already works very well. In contrast, matters concerning the church are not what they should be. In that respect, people here are very narrow-minded. That is terrible because the church here is very beautiful, but the condition it is in is awful. This is proof that in terms of religion, absolutely nothing was done here. From what my boys tell me, three-fourths of them (i.e., of the people in the village) don't go to church. They all left the church. Communism is responsible for this, and thus one can't expect any solidarity [based on religious ties]. For example, for us it is self-evident that a farmer should attend the church service for Thanksgiving, but none of the LPG managers appeared. I also find it natural to attend church with my children for Christmas, and from time to time in between, for example, on the anniversary of the death of a family member. I am extremely disappointed in this respect that no one is prepared to do anything at all. Consequently, people who have come here from the west have had to take matters into their own hands and have decided to do something for the church. First, we bought a few windows. The rain used to trickle down onto the altar. No one seemed to care."

Similarly, the nobleman also contributed to associational life and to the church by joining the Rotary Club and paying for church windows. He had originally come with many ideas for local development, but had given up on them because of the struggle to obtain land from the cooperative. He seemed to have contradictory feelings about working with others. On the one hand, he said that he had to act alone rather than in consort with others, but on the other, he believed that working together with cooperative managers and others in the Rotary Club to establish a kindergarten and a meeting place for young people had led to an atmosphere of camaraderie.

Finally, our oldest western informant won the hearts of his adoptive community by throwing a large inaugural party (he used the term *Einstand unserer Familie*, which means "entrance into an office or privilege of the family") for the day of German reunification and Thanksgiving attended by the whole village and surroundings, altogether some 280 persons. "We had just finished the new barn and the pastor, a woman, held the service there. I am the chairman of the trombone club, so they came with their instruments and so did the brass band. We served a cold buffet, people could drink as much as they wanted. That was really wonderful. Now they are waiting for a repeat performance. Last year we didn't do it, but this year we want to hold a village feast again before the harvest. This year I will become 65 years old, another reason to celebrate." The villagers simply referred to our informant as 'The Farmer.' He commented, "I am the only farmer in this village and so they say, 'Couldn't 'The Farmer' contribute to this or that?'"

This informant and his wife were involved in the community in other ways as well. Having obtained a permit to hunt on the land he rented for the next twelve years, he was able to continue a hobby he had practiced in the west. He became the president of the hunting club. He and his wife both participated in the local choir. Finally, he was also a member of the council of elders of the producer's cooperative. To be sure, this was a far cry from his involvement in civic affairs in the west. "There, I was rarely at home in the evening; perhaps only one evening a week. Otherwise I was constantly on the road. I had thirteen official positions: town councilor, member of the local church board and the district church board, president of the trombone club, and so on. It was quite something. Well, I want to have some peace now. There isn't too much going on here in terms of voluntary associations anyway. There is only the fireman's association. They meet every two weeks in the fire station, but that's about it." As president of the trombone club, he regarded it as his major task to recruit teenagers in order to get them involved in worthwhile activities, away from alcohol and out of mischief. Except for the fact that state-sponsored activities had fallen away, he didn't see any differences with the west. People even sang the same songs. But his wife commented, "When they are in a joyful mood they will sing some of their old battle songs, also Russian ones. People from Lower Saxony are a bit more dour than they are in Sachsen-Anhalt. The latter are quite lively. They talk fast." Her husband agreed. "When you observe women here, you notice that they are fast talkers. Women have the say in the family. They dominate here. They also know how to talk. One can barely follow what they are saying. One has to listen carefully to understand. But it's fun."

The nobleman's solution to becoming accepted was to deemphasize status differences. He explained: "To answer your question about how the *Junker* and the cooperatives get along, you must understand that they don't regard us as *Junker*. I sit on the tractor and work alongside with the others. I shovel sand and am just as covered with dirt as my people. I do, however, feel a sense of responsibility toward my ancestors. That is one of the reasons for this bold venture. I get a kick out of farming here, exerting myself, and working with my neighbors."

As we have indicated earlier, by 1999, our western informants had made the transition in their new lives in the east quite smoothly both economically and socially. In particular, their children seem well integrated in their new communities. And yet, differences remain. Thus the administrator for the nobleman joined a group of businessmen who discussed village issues, such as the establishment of a Christmas fair. "I tried that for five years," Mr. L. commented disappointedly, "but the interest of the other members is limited. We are lacking a middle class here. The circle of private friends is very limited outside agriculture. I am sure that things will be better for our children."

The demesne manager was more explicit in his descriptions of the process of making a new life in the east. When asked whether he had been able to establish new contacts he answered, "Yes, we intensified contacts, we are establishing a new circle of acquaintances through our children. We are getting to know all sorts of interesting people. And some of the acquaintances have become friends. We know the people well and know how to judge them." When we suggested that perhaps the years of turmoil also helped in intensifying interaction, he agreed, and told us about a group of friends centered in Bernburg, describing it as "a completely new start, a chance to do something new. I am sure that the pioneering contributed in binding people together, also across occupational groups. The leader of that circle of acquaintances and friends is a humble man and places a lot of value on diversifying it. I am thankful to be included. This circle organizes such things as small exhibitions. One of the leader's friends is an artist. He invites friends and acquaintances to come and tries to open the political direction of the circle, because some of the group members, particularly the entrepreneurs and employees, tend toward the right. He deliberately includes people who are oriented towards the *Partei des demokratischen Sozialismas* (PDS) (the successors of the GDR Communist Party). In this way, we build bridges and become more self-critical. Constant consent is not a good thing. . . . Seven families belong to this circle. I am the only farmer. Five of us are from the west." Ultimately, this consultant feels that his and his wife's circle of acquaintances includes both easterners and westerners, but that all their close friends are still migrants from the west like themselves.

While the demesne farmer and his wife feel that in many ways the east has become more like the west, in other aspects of their life, they feel that they differ from the mainstream middle-class easterners. In terms of consumption, things have equalized. On the other hand, she sees as a major difference in the fact that women either work or send their children to day care centers even if they have lost their jobs. "That goes against my beliefs," the wife said. "But I don't discuss that any longer with them. As one eastern German woman told me once, 'We have been socialized differently.' I have adopted that saying myself. There is no point in discussing the issue further. There is no common ground. The same thing continues in school, in preschool. They send them there for the entire day. And we stand out because we don't participate. Out of thirty-four children of that age, twenty-two go to preschool. Those who don't go are all antisocial. The parents don't have the means to send them. As a result, we sometimes wonder whether we should send our children to preschool ourselves, because they would be with children who are socially more secure. As things are, they play with the ten children who don't go to preschool. Those are the conflicts. But I think that the friends we have found among the eastern Germans have a similar attitude as we do or maintain a certain tolerance."

The demesne farmer also continues to support church activities and has recently become part of the Protestant church board. Since the local congregation in his village, with a population of 400, is too small to warrant continuing its own church, they have had to join forces with neighboring villages, at the cost of local solidarity. There are only two children from households in the village who are churchgoers and their three children were the only ones who had been baptized in the past six years, so regional groupings would be important to maintain a Christian tradition. He feels that the church should go out and recruit, but is hesitant to take on such a project on his own. "I don't know how I could go about it. If I had someone else of the same age as I am, someone with enthusiasm and also with children, I could do it. But I don't feel in the position to do anything alone, as an outsider." He also belongs to a voluntary Christian association that is comprised of 50% easterners and 50% westerners.

Mr. Rascher feels that the adaptation has been mutual. His workers have adapted to his rhythm and he has become more sensitive to them. As he put it, "There have been a lot of changes—mutual adaptation. Work no longer stops at 4 P.M. and no longer begins at 6 A.M. but only at 7:30. The workers have become very flexible in terms of the way in which they work. Things certainly have changed for them, but they have adapted quickly and, after all these years, they have become used to the new routine. We too had to adapt as newcomers, especially in terms of the way we treat people. We can judge them better. There has been some growing together after all. It is certainly still difficult to establish close friendships. That takes time. If one moved within the old *Bundesländer* it would go more quickly. One reaches the same level more rapidly. Well, it is true, too, that not everybody is an entrepreneur here. That is certainly a factor too. But the fear of contact (*Berührungsängste*) is still there. . . . We are happy that we have been able to establish friendships with people here, but as far as fear of contact is concerned, you can't imagine the kinds of problems the eastern Germans still have about it. They still ask right away whether we are from here, or say 'You are from there,' things like that. It hurts. The wall is still in the heads of the people. . . . The children have the least problems with it. They go to the kindergarten and have a lot of fun there. And the parents are trying to get closer (*sich beschnuppern* [like animals that sniff each other]). We too have sought to establish contacts, many with people from the old *Bundesländer* who have established themselves here. My wife also has some eastern colleagues with whom she has established quite deep friendships. In the meantime things have become more fuzzy."

Mr. Rascher also feels that he and his wife have become sensitized to issues that strain relationships. For example, many degrees given before or at the time of the *Wende* were no longer recognized, and so people became disgruntled as a result of being demoted. The newcomer is therefore looked at warily "by some of the colleagues, not necessarily by the average. The problem is the high unemployment rate. Here, in the former GDR, people identified with their jobs. Everybody worked. Women brought their children to day care centers when they were four months old. They had their work collectives. And, after the *Wende*, all of this was questioned. Many a person has suffered. The social peace has been eroded. Many have succeeded, drive a large car, and can renovate their homes. Others were left by the wayside. The older ones, 45 to 50 years and up, did not have the opportunity to retool. They could not keep up and are unhappy. The situation is mirrored in election results. The PDS has responded to these feelings and has gained support. Things are boiling here. It is a dangerous situation. The PDS has incited it further. That is not the proper way to do things. People have to understand the market economy. They should be educated about why all these changes have occurred." The fear of being seen as too ostentatious and of attracting the jealousy of the neighbors has made Mr. Rascher uneasy about driving his recently acquired SUV, which has replaced his aging small car, a discomfort we ourselves felt driving a new car we had purchased for shipment to the United States at the end of our stay. The greatest pain, especially for Mrs. Rascher, was the fact that easterners would still criticize westerners in front of her, adding, of course, that she was not like the rest when she pointed out that she was a westerner too.

Like the demesne farmer, Mr. Rascher became involved in community affairs. Indeed, he ran for a position on the village council and served for several years. "It was a sign for me that I had been accepted into the community," he explained. "I ran for office to find out what kind of response (*Resonanz*) there would be to a newcomer, a Wessi. . . . It became too much for me because I was subjected to a lot of criticism. I also didn't have the time. . . . I cannot say, [however, that people would not listen to me because I was a westerner]. At first, I felt rejected by the mayor. . . . There was conflict there. . . . I also didn't have a lot of experience. My father gave me many tips. In spite of this, we were a good team. It was also fun. . . . It is important for young people to be on the council. They don't have to be westerners, though. Perhaps a westerner can avoid the mistakes that have been made back home, but the question is whether that knowledge can really be transferred." On the other hand, unlike other western farmers, Mr. Rascher has not taken on offices in the voluntary associations that have developed since the *Wende*, although he belongs to the fishing club, likes to go on communal hunts, and helps sponsor both the soccer club and the firemen's association. Finally, two of the westerners have joined the Rotary Club which includes more easterners than westerners (but, significantly, no eastern farmers).

In conclusion, the western farmers are solidifying their economic position in the east by leasing more land and buying as much as they can. In one instance, the purchase of land will reinstate the private ownership of large landed estates in the former GDR. Already in 1993–94, these farmers were intent not only to establish an economic foothold in the east but to establish a long-term presence, a presence that their *Wiedereinrichter* counterparts could take for granted even if they or their families had moved to the area relatively recently. Some could justify that presence by evoking historical claims; others could argue that they had been invited by a relative; still others saw themselves as stepping into well-established roles, such as demesne lessee, a role that traditionally gave a family a long-term claim to a farm; finally, they could legitimate their presence by assuming traditional obligations vis-à-vis the community such as sponsoring feasts and assuming leadership roles. This effort at becoming part of the landscape has itself evolved over the years. In our interviews, Mr. Rascher's emphasis has shifted from his kin ties to the region to his involvement in local affairs. And, as his children were beginning to develop a sense of who they were, the demesne farmer has begun to stress his family's long-term involvement in agriculture, seeing in his 5-year-old son as a potential successor. "Recently, he told me that he didn't want to go away from here but wanted to stay on this farm for ever. One does notice that my ancestors, who didn't do anything else for 400 years, are in all of us. He also only wants to study here. I can see what it means to be bound to a locality, to have a *Heimat*, a place where one belongs. This is why, while many others went to America, we stayed here. His sister is very different. She wants to go on vacation and experience the modern. When the children talk to each other, then H. wants to be a farmer and A. the farmer's wife. (When she plays with a girlfriend she wants to be a nurse.) They also take on these roles when they play. Sometimes I myself find it too much. But they see us in these roles." On the other hand, this consultant's insecurity about his long-term position in the east is manifested in the fact that he is careful to point out to his son that "We are not owners of this land, so that unrealistic expectations are not instilled."

Even the farm manager was thinking not just about the farm's long-term future but also about his own family's stake in his boss's farming ventures. He hopes that one of his sons will someday succeed him and his other son perhaps take over the management of one of his employer's other farms in the west.

These western "strangers," then, have become familiarized with the new situation in the east. They are still sensitive to perceived sleights but are more conscious of their neighbors' sensitivities too. They don't want to be classified together with the stereotypical ugly, rapacious west German who has no long-term commitment to the east. So they are wounded when their overtures are rebuffed. They feel that they have paid their dues, as it were, by suffering through the transition and should by now be fully accepted in the community. A slow, mutual adaptation is taking place but they are not close; for they identify with a system that remains threatening to their neighbors, many of whom have not benefited or prospered in the reunified Germany. Since their children do feel "at home" and even grandparents have been imported, they hope to establish long-term roots based not only on property but on civic engagement too. Their prospects rest on the reduction of economic disparities between the two Germanies and a better appreciation of their different histories.

CHAPTER NINE

The Service/Industrial Context of Agriculture: Linkages and Permutations

In the previous chapters we have dealt with the relations of the various farmers to one another and with the state and its agencies, the institutions that aid their endeavors like banks for capital and major suppliers and outlets, the educational system for apprentices, and the communities that surround them. But the successes and failures of the farmers of Sachsen-Anhalt cannot be understood without a more focused analysis of the service/industrial context in which they are embedded, that is, the suppliers, service personnel, and clients with whom the farmers are in direct contact and upon whom they depend. The convergent and divergent nature of the development of these firms, which contribute indirectly to the viability of agriculture and influence its direction, will highlight the special nature of the farmers we have described and analyzed.

Suppliers

As a basis for our discussion of the transformations after 1989 from the vantage point of the suppliers, we will use the examples of a medium-sized eastern German agricultural equipment dealer and small-scale manufacturer, a large-scale western German tractor dealer, and a small-scale electrical firm. Two cases, one of a very small-scale eastern German miller and another of a large-scale western German one, will illustrate the transformations from the vantage point of the clients. While it would be impossible to generalize from their experiences, they do illustrate some of the adaptive mechanisms employed by eastern and western firms, many of which were also employed by the entrepreneurs we interviewed in other sectors of the economy as well.

Underlying the success of both tractor and agricultural equipment dealers was the fertility of the region and the large expanses of agricultural land there. In addition, the short-term need for the eastern farmers to catch up with western technology provided a means to jump-start the enterprises. Our first example is of Mr. A., an eastern German entrepreneur who sold trucks and farm equipment and also produced conveyor systems and equipped the trucks he sold with custom-made loading devices. Mr. A.'s dual training in international commerce and machine engineering and his work in both fields (he first worked as an engineer in refrigeration and then, for over thirty years, in the export department of a factory that manufactured agricultural machinery) made him well suited to undertake the move toward establishing an independent firm. His experience as a representative for his firm all over Europe as well as in Africa and South America also came in handy, for it provided him with business connections in the west, broadened his horizons, and put him in a position with a greater degree of autonomy. The year before reunification, he had already left the old firm and tried to establish his own workshop, but was stymied because during that phase no financing could be obtained. Then he decided to sell machinery on commission. He commented, "Because of the contacts and partners I had in West Germany, I found it easy to say to myself that I would take goods on commission and would sell them. Agriculture in this area is highly productive. We have the best possible soil structure and when there is enough rain, everything grows. In fact, it would be impossible for me to give up the sale of agricultural machinery. In other words, my strategy of beginning with agricultural machinery was quite deliberate."

Rather than taking over a section of an old state firm, which would have entailed firing workers and dealing with the firm's debts or pollution problems, Mr. A. decided to begin from scratch with his wife holding a small share of a partnership. "I started out with my Lada (Russian-made car) and sought business partners, or rather individuals whom I imagined might have that potential. It was a lot of fun, but I did it more on a whim rather than anything else. It wasn't far, so one day I went to see the N factory. So when I went to try to obtain the contract I drove to X (in western Germany). I knew how to get there but I didn't know the streets. So I arrived long after the time for my appointment that I had finally succeeded in setting up after half a year of trying. All the members of the sales department had already left because I arrived around 5 and my appointment was for 1 in the afternoon. I was just able to catch the head manager. He offered me a cup of coffee, but I told him that I had not really come to drink coffee with him but to ask him for a contract.

"I was wearing jeans and a pullover and you can well imagine the laughter that must have occasioned. I found out only later that I had been talking to the head sales manager in a hierarchy where it would have been appropriate for me to wear jacket and tie. But they have excused my mistake. That was just to make the first contact. The idea was not just to build up the contract in a single-minded manner, but to create a human relationship. It was only weeks later that I asked them whether they would not at least give working with us a try. I told them that I knew what farmers needed here. The first order of importance was to build milking systems. We would buy them and install them. For milk would generate the first cash flow for the farmer. We would concentrate fully on the barn, the tractors could come later." Soon thereafter, Mr. A. branched out into other activities. "When a new task appeared, I would hire good people for it and so we added one activity after another. It was necessary to continue in order not to lose ground and so after agricultural equipment came trucks and equipment for public works, in other words, an incredibly broad spectrum, all in the first two years."

Since then, Mr. A. has had to reduce the scope of his activities. He emphasized service over production, even though he felt that it was a shame not to be able to take full advantage of the many trained workers. For the labor costs compared to other countries farther east were prohibitive. Nevertheless, the services he provided included the installation, adaptation, and even the partial production of specialty equipment. For the latter purpose he purchased a large hangar near Halle to custom make truck parts such as crane supports and tip-truck superstructures and even conveyors which he joined to standard elements.

For this versatile entrepreneur, the *Wende* opened up a world of opportunities: the realization of previously unfathomed potentials of the multiple skills he acquired over a lifetime of work in GDR factories and the liberation from restrictions on his freedom to express his ideas and travel, particularly with other family members. On the other hand, the excesses of capitalism and the reorientation of values also gave him pause. Even though he himself was able to travel widely before the *Wende*, he was never allowed to take family members along with him and so he was not only isolated from them physically for three-month stints abroad, but they could only participate vicariously in a major part of his life. Also, just when he was beginning to establish personal contacts among his business partners abroad, he would be called home and later sent elsewhere. Thus, in Nigeria he found the people very receptive to new ideas compared with other countries he had visited. "Even though the conditions there were less than optimal, they managed well with the little they had. But such a stint would only last a certain time and then [the authorities] became uncomfortable about the contacts I had established. They were too good. So back home I went. . . . There were many smartboys (sic, i.e., spies) with us. One could distinguish them by the fact that they did not know how to work. They had a good education but they had no experience, which was quite obvious. So they would be sent out to organize dinner and fetch beer. They would also establish contacts. It was part of the game to put a good face on a bad situation. You can't possibly understand our desire then to experience something different, whether through sport or through professional opportunities, opportunities that few enjoyed." He himself had belonged to the Communist Party, which recruited him because of his involvement in competitive sports. However, a year before the *Wende*, he quit the party, was relieved of his managerial position, and his income was reduced by more than 10%, not to mention the fact that he became the brunt of the opprobrium of his fellow workers.

But Mr. A. was also critical of the new era. Ironically, now that he was free to travel with his family, he had no time for travel and felt that he would have to postpone showing them the places he had known until his retirement. He also regretted not having the time to read a book or go to the theater, something that had been more possible as an employee in the GDR. This nostalgia for a more relaxed pace was an often repeated refrain among former GDR farmers and other entrepreneurs.

Mr. A., then, combined his unusually versatile educational background and work experiences with new opportunities: access to capital, a highly trained workforce, a sense of providing services as an avocation, a deep knowledge of agriculture and the limits imposed on this sector, and a tradition of bricolage characteristic of an economy of scarcity. As we have also seen in the chapters on the collectives managers this required not only making do with whatever was at hand, but gaining access to inputs by cultivating personal contacts, and a camaraderie among workers facilitated by the lack of marked economic distinctions. Mr. A. was in the process of establishing a family business that he hoped to pass on to his two teenage sons who had already expressed an interest in joining. The danger, perhaps, was that he may have been overextending himself. Indeed, by 2001, only one of his enterprises was still operating.

In his dealings with farmers, Mr. A. emphasized the importance of establishing long-term relationships with his clients and providing advice with the goal to maximize their long-term survival rather than short-term profit for his own firm. The fact that he offered everything from dairy equipment to tractors and combines facilitated this strategy, a strategy that he differentiated from the one he employed with his other clients, the transportation departments of construction firms. "I know all the farmers here, I know almost all my clients, including those who relocated here from elsewhere. The farmers often pay me a visit for they are curious to find out what is available now for agriculture after making do with so little in the past."

Our second example, the case of a western German distributor of agricultural machinery, has many parallels with the previous one, not only because the firm is located in the same fertile area (in fact, in the most fertile subzone of the region), but also because it was established thanks to the collaboration with an eastern German manager who received the encouragement from the owner to put his imprint on many aspects of the operation. This manager, Mr. S., whose experiences during the GDR were the subject of our earlier discussion of farm equipment procurement and maintenance under communism (see chapter 4), had worked as the head of the farm equipment division of a large model state farm and upon the *Wende* was, as we have discussed in chapter 6, encouraged and given time off from work to establish contacts in the west. After an initial failed attempt to continue the tractor operation as a large repair shop and then as a large and unwieldy repair shop cum tractor dealership, Mr. S. and his colleagues decided to search for someone to take it over as a smaller operation. Unable to gain access to capital to start his own business, Mr. S. sought out a western entrepreneur with greater financial means. That search ended just one month after the border opened, when he found Mr. B., a major intermediary for a North American tractor company. By March 1990, Mr. B. had already staged a show attended by 2,000 persons on the premises of the former state farm and, spurred by the enthusiastic reception, rented and later purchased one of the well-kept buildings of the model state farm. This early start earned him an exclusive distributorship. He considered himself very lucky that there were no prior claims to the property and that he could construct a new building for the storage of tractor parts, quite rapidly.

"In the beginning, we had enormous problems," he explained. "We didn't even have a telephone. Then we had one. But it often failed. We had to place the order for parts by phone and so twenty people would fight for the use of the line. We now have some eight to ten telephone lines, which works out well. We started with only one computer. Much later we had a full network and now working conditions are quite good."

Mr. B. and his father's life entailed a lot of commuting between different locations. "I live on the Autobahn between H., [one and a half to one and three-quarter hours away] and here. I am constantly commuting. We have built a house here where all of us stay when we come from over there (*drüben*), that is, also my father and someone who takes care of the Czech Republic from here and, finally, a former employee of the factory who stays here three days a week."

Since Mr. B. and his father continued to run a distributorship in western Germany, much of the day-to-day operations of the firm lay in the hands of his eastern German colleague. Unlike Mr. A., his east German colleague who could build on his past experience, Mr. S., the eastern German submanager, thought that the most difficult aspect of adapting to the post-*Wende* situation was the fact that a large part of the knowledge that he had acquired during the fifteen years that he had worked for the state farm had become obsolete. "All the new things for which I had expended a lot of time and energy to understand, such as new regulations and technologies as well as regulations regarding the protection of workers, health, the manner in which welding had to be done, and so on, all became history with unification. This meant that in many aspects [of work] I had to start from scratch." Another change that took him time to get used to was the fact that he now had to make many decisions on his own without clearing them with his boss who was away most of the time visiting clients and drumming up new business. "When I came home in the evening my head would often spin. Before, I was constantly limited in making decisions. We had always been told, 'This is none of your business. Take care of your own business.' Now I had to think about everything, and simultaneously at that. When I had to make decisions about buying something or spending on advertising which always means an expenditure of several thousand marks, how would I know whether I had made the right decision? Before, I was relieved of the responsibility of making decisions. Whereas then, it was always wrong if one made any decision, now, it's wrong if one *does not* make decisions. At the time, any plan was first discussed to death, then it had to be approved, whereupon the money had to be requested, which would always be granted only partially. Finally, one had to see what one could do without and see what machines were actually available at the moment and so on." It took Mr. S. two years to become fully comfortable with the new responsibilities and opportunities.

It seems to us that Mr. S. underestimated his contribution to the firm. His knowledge of the farmers in the area was not obsolete. To establish a clientele, he immediately resorted to his contacts with the heads of agricultural cooperatives and with private farmers who had once worked for the state farm and had established independent farms. "In the first period, [my contacts in the area were especially useful] because I knew everybody. There was a strong sense of togetherness. I knew exactly when I called a person in a neighboring LPG, what his wife's name was, how many children he had, what his family life was like, and so on, because we had visited one another after work and seen each other during weeklong training courses. So the contacts were there and they came in very handy. Now they are slowly falling apart because these men no longer work for the old farms but work elsewhere, often in unrelated fields, and so we are growing apart. [But then] we knew who had taken over what land; what they leased and how much they owned and therefore what machines they could use.

"I could also make concrete offers beyond the machines to spare parts, additional equipment, and so on. I knew who could act as a supplier for such specialty items as hydraulic elements and the repair of electric equipment such as starter motors or dynamos. The companies that produced these items were not in the area and one couldn't take a trip to Braunschweig for a 10 mark item. If one knew that such a person had privatized in such and such a direction, or worked in a company that carried a particular item or provided a particular service, one could save time and money." The relationship with clients continued to be highly personal. "We know one another and we charge fair prices. Also, and most important, if they have a problem on a Saturday or Sunday, they will come to my home and say, 'I need this or that.' Then I take them here and they can use the workshop when they want to do something themselves or they can get the parts from the storeroom. We help one another."

Unlike Mr. A. in the first example, Mr. B. carried only tractors and combines, which he also sold directly within a radius of some 35 to 40 kilometers and through a network of some thirty smaller retailers located within 100 kilometers around the distributorship. All these dealerships, ranging in size from three- to twenty-worker operations, were owned by eastern Germans. With great foresight, Mr. B. was able to predict the process of privatization of machine workshops. He described the establishment of these dealerships as follows: "Among the first tasks I and also my father undertook when we started here in 1990 was to visit our clients, the LPGs, and we also tried to convince the managers to take over the workshops because we believed that the farms would remain large, 5,000 or 6,000 hectares, and that farms of this size would require workshops of their own, even if on a smaller scale than before. We believed that such a shop could also be privatized [separately], and so we encouraged [the workshop managers] to establish private firms and tend to the machinery pool of the LPG as well as the machinery of other farms in the area. That is how we established contact with these people. Those were all technical managers of the farms who would have become unemployed if the LPG decided to discontinue its own work shop. At first, some LPGs continued running a workshop as an affiliated company (GmbH). But, since then, the workshops have all become private."

The year 1991 produced spectacular results when all the cooperatives as well as the newly established private farmers sought new equipment. The distributorship was particularly successful, since it was able to establish its own leasing or lease/purchase program by borrowing money directly from a bank rather than through a leasing company, and because it was able to order sufficient agricultural machinery from the United States to cover the burgeoning demand. Soon they had gone over the initial 10 million deutsche marks credit line and there seemed to be no end in sight, for renting a combine enabled a farmer to keep the costs per hectare to under 100 deutsche marks, which was substantially less than it cost to hire a specialized firm to bring in the harvest. After a few years of use, many of the leased machines were, as we have already indicated, sold in western Germany, where small-scale farmers were attracted by the lower prices. The expenses incurred by work stoppages resulting from repairs in their smaller operations were not as deleterious as in the large operations in the east. Others were shipped to the Czech Republic, Greece, and as far away as Syria. At the time of our first visit, the eastern German operation contributed almost half of the firm's total annual turnover of some 46 million deutsche marks. The branches employed around thirty persons each.

Mr. B. found it easy to retrain the mechanics hired for his repair shop to deal with western technology. "We noticed right from the beginning that the understanding of technology was very good. Precisely because one had to improvise a lot in the past, it was no problem to confront the mechanics with the new technology. That was really good. Of course, the electronics and special hydraulics posed new challenges for the mechanics, but in the last analysis, the technological understanding was there. We sent a few of the mechanics to the mother firm in H., and (the manufacturer) also has a training center [in the west]. The important thing was to give them an understanding of the work routine in our workshop [in the west]. And so all of us learned something in the shortest order. Time also took care of many things, because one can't teach everything. It was a very rapid but continuous development. Especially in the beginning, the fact that we had recourse to the workshop in H. was a big plus. We could phone at any time and ask the master there what to do under such and such circumstances and he would often be able to give us a diagnosis right over the phone, or, when the problem was too intractable, he would tell us to disassemble the machine and wait until he came." Mr. S. added, "That was the time when he was often here anyway and he would help out. So we would undertake three or four special tasks on that day and he would tell our people how to do them and so when he returned home in the evening our people had made more progress than if they had taken a course for an entire week." The firm also provided training sessions for the affiliated dealerships and they also received training at the factory in the west.

The firm also had two or three combines on hand that they could loan to farmers when their machines were being repaired. The large scale of the farms, the fact, for example, that the largest client, a cooperative that farmed over 16,000 hectares of land (including land that it cultivated for private individuals), had thirty of B.'s tractors and fifteen combines made it worth while to go the extra mile to keep the clients satisfied.

In sum, the foregoing example illustrates the potential synergy between a western German capitalist and an eastern German submanager. Mr. S. was given considerable leeway to impose his style of customer relationships on the firm. The rapid expansion of the firm into the regional market would not have been possible for the western entrepreneur without the ready-made network of ties established through decades of informal exchanges among state firms and LPGs. In the GDR, these ties had emerged out of the need to counterbalance a rigid system of allocation within and partly resulting from an economy of scarcity. That system of allocation of course also put stringent limits on such exchanges. With the demise of the system but the ties still in place, the latter were employed to new ends even in the initial restructuring phase. By the time Mr. B. arrived on the scene, the network already operated in the commercialization of farm equipment. The westerner's capital and the well-established supply line to western technology gave an impetus to the full utilization of the network, although many of the personal ties were already falling apart because the hectic post-*Wende* life, the dislocation of many of the actors from their positions of power, the fact that emerging differences in the standard of living reduced face-to-face interaction, and, last but not least, the expansion of the firm beyond the geographical limits of the original social network.

As in the previous example, the flexibility of the mechanics and technicians to make do with whatever was at hand translated into flexibility in learning new techniques, although, by 2001, apprentices were no longer versed in the skills to repair rather than replace parts. Flexibility was greatly enhanced by access to western expertise that was never farther away than a short visit from a mechanic located at the parent company or even just a telephone call. In the process, the eastern German submanager was empowered to continue to act on his own as he had during the transition period, but he also had to learn an entirely new way of stocking parts, streamlining service to correspond to the newly competitive environment, learning a new system of financing, and coping with a telephone network that was not yet geared toward the new need for rapid communication.

Our visit of Mr. B.'s dealership in June 2001 confirmed our earlier impression of an exceptionally vibrant enterprise. They were able to expand in spite of a generally stagnant market for agricultural equipment—Mr. B. estimates that it has shrunk by one third since 1994. True enough, the sale of large numbers has ended, although the machinery has been replaced once since the *Wende*. Leasing has become more prevalent because the farms are now investing in land to prevent others from purchasing the land that landowners and the *Treuhand* are increasingly offering for sale. In the addition, the farmers must now begin to amortize loans and no longer enjoy their earlier tax holidays. Therefore, farmers have little spare cash at their disposal. Sales have also decreased because farmers were initially able to take advantage of special generous depreciation allowances that permitted them to reduce taxable income when they invested in machinery. Since rental expenses can be deducted from profit, lease agreements have become more attractive. Finally, some farmers now prefer contracts that include maintenance, thereby obviating the need for mechanics. But the firm was able to increase its market share and therefore did not suffer as much as other dealerships. The market for used tractors, sold after the lease period is over, changes constantly. Thus different eastern European countries became major clients only to be eclipsed by their neighbors in subsequent years. The firm's location in the middle of Europe is a major advantage, for it provides privileged access to all potential markets in Europe. Mr. B. predicts that the demands on tractor dealerships all over the world will become increasingly onerous, with manufacturers assuming fewer and fewer responsibilities. As things are, the profit margin on agricultural equipment is only 0.3% of its price. With the possibility of direct internet sales, "Eventually, the manufacturers will only pay a certain sum for specific services provided by the dealers. Then, the art of [remaining afloat] will consist in becoming more and more cost efficient and streamlined." For example, in the future satellite positioning systems will enable dealers "be able to see how many hours a tractor has been in use and when it needs to be checked and we can tell the client when they will have time to do it." They win also be able to send messages to the farmer and the workshop of breakdowns including the location of the machine and the nature of the problem. As a first step, satellite positioning systems to measure the characteristics of each field are beginning to be introduced into the area. They have been installed on 30 of the 180 combines sold or leased by Mr. B. during the 2000/2001 season, up from 7 in 1999, when he first offered them for sale. They cost 20,000 per combine, or 6,000 for their annual lease, adding up to 12% to their basic price. While their cost effectiveness is not yet proven, the largest farms have decided to become pioneers in their use. Mr. B. foresees a few years of data gathering before farmers take any further steps.

As in many other instances of west-east cooperation within western-owned firms we investigated during the broader study undertaken in the region of Bernburg, we were impressed by the apparently smooth functioning of the relationship (for other examples, see Buechler & Buechler 1999a). The case belies the eastern stereotype of the westerner as a bully, showing, instead, an easy relationship and the western German's sensitivity to the need for a personal, humane relationship. We observed many of our western informants who rejected western stereotypes of the easterners and stressed the importance of observing a different management style. The interaction between Mr. S. and Mr. B., and the manner in which each permitted the other to explain the part of the story with which he was most familiar, gave the impression of a team in which the partners appreciated their respective strengths. Mr. B.'s willingness to become integrated into the region has certainly contributed to the harmonious work climate in the firm. By 2001, he had married a woman from nearby Magdeburg and feels fully at home in the region. The fact that he has spent one third of his life in the east has helped him to sever his ties to the west. In addition he claimed that "The mentality of the people in Niedersachsen is similar to that in the Magdeburger Börde, they are both influenced by their involvement in agriculture. The forty years of communism did not change that. . . . There is no east-west conflict here. It has been a long time and the relationship is regarded as normal." "The people who came from the west and were unhappy here have gone back again and so it has become a, homogeneous system again," added Mr. S., to which Mr. B. replied, "We can all laugh at Wessi jokes and when there is a film about the GDR we can become involved too." While Mr. B.'s example shows the fallacy of the stereotype of the pushy westerner, his rapid reaction to the opportunities did gain him an exclusive distributorship over a vast highly desirable region. The case also demonstrates the dynamics of the pecking order in the formation of enterprises that is particularly in evidence in large-scale commerce and industry, with the westerner commanding access to considerable resources on top and the eastern German in sometimes responsible but at the same time usually subordinate positions as an employee or local dealer. Such subordinate positions were rejected by our eastern German cooperative managers and private farmers and also by the westerners, even the farm manager.

A Serviceman

Our final example, a service enterprise related to agriculture, complements the picture provided by the first two cases in that it replicates the trajectory of the entrepreneur in our first example but on a much smaller scale. It also echoes—but in an inverted manner—the experiences of Mr. S., the eastern German manager in our second example who was prevented from establishing his own enterprise due to the lack of capital. Like Mr. S., Mr. J., the owner of this enterprise, had also worked in a service division of a large farm, but he waited until he lost his job before looking around for ways of earning a living. He would have preferred to continue working as an employee, but found that establishing his own enterprise was the only way he could secure a living.

Mr. J. worked as an electrician on a state-run seed farm for twenty-five years. His career was intricately linked to the dynamics of the GDR system. As he explained, "In GDR times, professional training was writ large. It was unheard of for a young person not to be able to get an apprenticeship position. So that system also had many advantages over the present one. Of course, every firm tried to train apprentices in professions where the cost was low. An electrician required only a bag with tools and a testing board. The result was that every other firm in Bernburg trained electricians." Although wages were better in industry, Mr. J. decided to stay with his job because of the fact that it encompassed a great variety of tasks, ranging from wiring large new seed-drying installations, to wiring offices and workshops, to repairing conveyor systems and cleaning equipment, and even pump stations that provided water for the communities associated with the farm, and, for a time, doing electrical work on farm vehicles. "[This variety] was not the case in large enterprises. There, each electrician would do only one type of work. One would deal only with motors, another with lighting. I was impressed by the variety. That's why I stayed for twenty-five years."

In addition to his regular work, he, like many if not most craftsmen in the GDR, was also engaged in providing services for individuals after hours. "Sometimes we would work for two or three hours [after regular work] for weeks on end, as many as we could manage physically, and then not at all. But it must have amounted to 20% of our income. We were indeed very active. We could also help a lot in this manner. Often there was no one else who could or would have done the work. Now there are so many businesses. All you have to do is place a few phone calls and three appear who want to do the job. That was not the case then." Finally, Mr. J. also was able to fix up his own apartment and later the house he bought by exchanging labor with other artisans. He bought some scarce resources from his state farm, while other items were procured through networks of contacts with individuals in other firms.

After the *Wende,* Mr. J. continued to work for the seed farm for another year in the hope that it would be able to survive in some form or other. However, the different parts of the large complex went their separate ways. The seed farm was taken over by a western German who reduced the workforce from 120 to twenty-two full-time workers, without any full-time electricians. "When I was given notice, what was I going to do? I was too old to find employment and I was too young for early retirement. To be frank, I felt insecure about becoming independent. . . . I was too reticent and fearful and didn't want to do it. But when I didn't have any other alternative, I did belatedly become independent.

"I had a title as master, but I had to take another master craftsman's course in Halle. . . . The firm for which I had worked and that had given me notice told me that they would help me as best they could. Many other firms just dropped their employees. In times of crisis everybody only thinks about himself, but that was not true in my case. They let me rent the workshop cheaply so I could launch my enterprise."

Starting an independent business as an electrician was not easy. Mr. J. believed that there were ninety-two electrical installation firms in the *Kreis* of Bernburg alone, a legacy of the overproduction of electricians in the GDR. "Out of these, at least 30% will fold sooner or later," he said wistfully, "and my goal is to provide a solid basis so that I am not going to be among them. I don't plan to be rich, that is foolish. I plan to continue my work. First and foremost, I will keep my customers happy so that if my costs should rise and I have to raise my prices, they will stay with me anyway." Mr. J.'s investments were relatively low (certainly one of the reasons for the establishment of so many competing electrical installation firms). He had an old Wartburg (high-end GDR production) station wagon, took over a GDR delivery truck that had received a complete overhaul, and purchased an additional car on credit for himself and equipment worth well over 10,000 deutsche marks. Since the purchasing power of the majority of his customers was limited, he kept his prices low. His cautious assessment of the future of his enterprise not withstanding, he worked with five workers and had more than enough work for everyone. Since he felt obligated to his workers, he preferred not to hire workers on a short-term basis except when he could borrow a worker from another enterprise experiencing a temporary slowdown. Three out of the five were known to him, having worked in the same work brigade. He chose one of his new workers because he liked the man's honesty and good will. The man told him that he had worked for a large state enterprise and had not had the opportunity of doing much more than running errands. He was eager to finally learn something.

Mr. J.'s versatility in his previous employment and his love of new challenges served him well. As he put it, "I was always a person who was attracted to complex jobs. I couldn't stand it, when something remained undone. When I couldn't do something during the first try because it was entirely new and strange, then I found out what to do and finished the task. That's how it is today too. So [my versatility] has certainly helped me. After twenty-seven years of working in the profession I am seeing things now that I never saw before. But if one is accustomed to variety, then one finds one's bearings more readily than someone who always did the same thing." He has also benefited from the courses organized by the electricians' guild and by the manufacturers. But versatility and the openness to new challenges were not Mr. J.'s only assets carried over from the past. He also continued to tend the electrical installations of the western German seed producer that took over his old firm. "Until now they are doing only essential maintenance. Nothing new has been added. In general, I have been commissioned to do the work. I know the installations so I can fix things in the time it would take someone else just to find his bearings. This has been to the firm's advantage, so I immediately had some jobs to do. When I go there after having known the installation for the past twenty years, I know exactly which screw to loosen and which circuitry board to open to find the problem. We would never hire three planners [to design the layout], as they do now, who earn money on every larger project. We would do it ourselves, so the circuits are [haphazardly] branched [rather than carefully laid out]. We would say what we needed. Then I would work out the best way to connect it to the existing electrical network. Then I would try to get the required materials and would meld the old with the materials that I could obtain. I would make my own drawings and we would build the installation and it would always work. If we had used engineering firms, like other companies did, we would have ended up with plans that called for materials that might not have been obtainable 30% of the time, and we could not have accomplished what we had set out to do. I combined theory and praxis and was the person on which everything hinged."

The responsibilities of a business of his own still weighed heavily on Mr. J. He had not taken a single vacation in two years and even on Saturdays he worked late on his accounts. "I can't sit still for a moment. When I am at home on Sunday afternoon and my wife tells me to sit down and relax, I don't know what to do with myself. My right hand looks for my left hand and vice versa. It's not a good sign, it shouldn't be that way." He said that he would like to teach his wife who used to work in commerce but lost her job after the *Wende* to do some of the accounting, but found that it slowed down the work if he had to explain the technical terminology, and so the hectic pace has impeded him from being able to share the burden of running his enterprise with her.

As in the previous case studies, Mr. J.s' example shows the carryover of the inventiveness that was necessary to come to terms with the rigidities of a redistributive economy stressed by most of our eastern informants. However, it goes further than this by showing how the extreme division of labor that characterized the Taylorist model that pervaded not only GDR industry but every other sector of the economy as well, restricted the means at the disposal of trained individuals to innovate. Each specialist was so narrowly trained that he or she could not readily have resorted to his or her training background as a source of innovation. Mr. J.'s rare versatility and his exceptional dedication to his work, which he regarded as an avocation, gave him an edge over his competitors. Such an exceptional advantage was particularly necessary in such fields as electrical work because of the overproduction of trained electricians that had no relationship to demand.

We were not able to contact Mr. J. in 1999. Although we left several telephone messages, he did not return our calls. In 2001 we had a short telephone conversation in which he told us that he was working with two trained workers and two apprentices. He said that he had personal problems and declined an interview.

Clients

Our next pair of examples represents the opposite side of the nexus of enterprises directly associated with agriculture: the customers for the farmers' output. The first example is a flour mill with a capacity of only 4 tons a day, not fully utilized at present, owned by the fifth and sixth generations of an eastern German family that operated it privately even during the GDR era. The second is a large mill—once a part of a state conglomerate that operated some thirteen mills—owned by a western German family enterprise that also owned two other medium-sized mills in the west. Both our informant who managed the mill and his wife came from miller families, his family in the second generation and his wife's for many generations. This mill was operating at the upper limits of its capacity of 8,000 tons per month.

The management of the small mill was taken over by the miller who ran it in 1994 from his father who died in 1966. The first document pertaining to the mill came from the eleventh century when the mill was owned by a count. Originally driven by a water wheel struck by water at the bottom of the wheel, the system was later improved so that the water would strike the wheel from the top. A dammed holding pond that could be filled during the night served to regulate and increase the water supply. In addition, the owner's grandfather still ground flour with a windmill nearby. Finally, sometime in the 1920s, an electric motor was added, which regularized the speed of the drive shaft, a system that was maintained until the late 1960s, when the residues dumped into the stream by a coal mine silted the holding pond completely, so that, by noon, the electric motor had to drag along the water wheel for the rest of the day, wasting electric energy. Then, once dry, the oak wheel began to rot away.

The state authorities during the communist period, who did not look kindly at the continuation of private enterprises, least of all productive enterprises, nevertheless allowed small-scale mills to continue in private hands, for millers were categorized as artisans rather than as factory owners. The rationale behind this decision may well have been the fact that such operations provided little competition and consumed few if any resources other than energy. Indeed, during the entire GDR period to the present day the mill was maintained by cannibalizing parts, purchased from small-scale millers who discontinued their operations. The danger of constituting a competitive threat to state industries was also held in check by strictly circumscribing the nature and scope of the activities a miller could pursue.

Our informant's specialty was rye flour, which he continued to produce after the *Wende*. "The milling of wheat allowed for a higher profit margin than rye, so we were allowed to mill rye while the state firms milled the wheat. However, we weren't even allowed to deliver our flour outside the *Kreis*, which was like selling to a foreign country for us. Today no one cares where we sell our flour." On the other hand, the original source of the grain was not necessarily local. He had to purchase all his rye from a particular state distributor and was unable to choose grain according to origin. The quantities were determined by quarterly and later yearly contracts, that were in fact upheld. "We never stood there without grain." And, to a large extent, the mill operated at full capacity. Our informant first operated the mill with his sister (an accountant in a neighboring LPG) and the help of his wife, an invalid who had retired from the agricultural cooperative. Although, in theory, the firm would have been allowed to employ other workers, our informant was never able to hire any. State enterprises paid significantly higher wages than those that private millers were allowed to offer, and so no one was interested in working for him. Shortly before the *Wende*, he was joined by his son-in-law, who was originally an electrician.

The *Wende* required few fundamental changes in the operation of the mill. Our informant cited competitive pressures from large western mills that were able to offer flour at dumping prices in order to increase their share of the market in the east and the fact that a large percentage of the private bakeries that continued to produce during the GDR have been closed as two of the factors that have lead to a decrease in demand for his flour. At the same time, he was not equipped with a specialized truck that would have permitted him to offer loose flour to the new western-owned large bakery established in Bernburg (see Buechler & Buechler 1999a). Unable to borrow capital, he was reluctant to invest in such a vehicle, or even in a container that could be mounted on their truck when needed. Also, most of his long-established clientele did not have the facilities to store loose flour. Indeed, the number of small-scale millers in Sachsen-Anhalt was dwindling rapidly because no willing successors could be found or because too many of their traditional clients had closed their bakeries. Our informant told us that in the past, some forty-five to fifty millers used to attend guild meetings in Sachsen-Anhalt. Now there were only ten or twelve. Mr. L. attributed the ultimate cause for the drop in demand for his flour to a decrease in bread consumption in general resulting from rampant unemployment. "We no longer produce the quantities of flour we did during GDR times. We can't get rid of them. It's logical. First and foremost, the consumption is no longer there, because we have so many unemployed here. I see it when I talk to my bakers. Recently one of them told me that his brother-in-law's family used to consume three [2 pound] loaves a week. Now they are both unemployed and consume only one loaf. So a lot of baking falls by the wayside. There is no need to work several shifts. It all adds up. That's just one example. Around here one will soon have to ask who *does* still have work. It is sad."

Mr. L. continued to maintain his operation by purchasing used equipment. The purchase of new, more energy-efficient machinery was impossible to justify financially even if the banks were willing to lend him the money. The irony did not escape him that, during the GDR, the small mills could not obtain any inputs and now, when everything would be available, they could not afford them. Mr. L.'s son-in-law continued to deliver all the flour to the bakers, which entailed unloading the

100 pound bags by hand. Hiring outsiders to transport the flour would have cut too deeply into the profit margin. The fact that this son-in-law was also a trained electrician further reduced maintenance costs of the fully automated plant.

To further reduce costs, Mr. L. was seriously considering reinstalling the old water wheel system. "There is a lot of movement in that direction. Just last week another commission came that wants to do something about the pond. Three months ago a larger commission was here and called me in to inquire whether I wanted to use the water again. I answered, 'At current energy prices, why not?' Now I have to wait and see what is going to come of it. The mechanic who works for me was, of course, immediately enthusiastic. He wanted a turbine to generate electricity rather than a water wheel. But the officials rejected the notion. Rather, if something is to be done, they want a water wheel again, the way it used to be. There is a bit of nostalgia in all of this." Reinstating water-driven milling would have entailed the dredging up of a lot of history both literally and figuratively. First, there was the problem of who would pay for dredging the residues of GDR mining in the dammed pond, for the successors to the mining operation denied any responsibility. Then, there was the fact that, at the end of the World War II, the allied forces dumped a truckload of mines they no longer had any use for into the pond that captured the outflow. Mr. L's sister feared that they were still dangerous and could explode when the pond was dredged. Other problems created by the past were solved on their own. The spring that fed the stream, which had failed when a nearby LPG dug a deep well to service its barns, reemerged in the two or three years since the barns were abandoned. The project, which would have received subsidies from the state, eager to promote clean sources of energy, would not, however, have reinstated the status quo ante entirely. The water wheel would have powered an electricity generator rather than the mill directly and the power would have been fed into the net.

While rye flour continued to be the specialty of the mill, other products had to be added. It began milling wheat into a number of flour types required to bake a new line of recently introduced bread types. Mr. L. was also trying to meet the rising demand for bran and rolled oats resulting from the increasing popularity of horses and ponies kept by the winners of German unification. To this end, he made contracts with neighboring farms for the production of particular grain varieties. For example, during GDR times, most rye came from the less fertile, sandy soil northern regions, and was only slowly being reintroduced in the Bernburg region. Mr. L. felt that the farmers in the region were still not sufficiently aware of the potential for growing and selling oats. Other ideas had not worked out, such as the sale of packaged flour to stores, which would have been economically feasible only if done on a very large scale.

Although the future of the mill was far from certain, its continuation for at least another generation seemed assured, if they could work out the succession. He told us that his grandfather continued to have a strong voice in running the mill until his death at 80 years of age and his father died of heart attack at work when he was 66. Legally he himself could have continued to earn without losing social security benefits after he reaching the age of 65. Mr. L. was in the process of negotiating his retirement with the authorities, but the paperwork for his retirement took far longer than expected. He had not discussed his role in the mill with his son-in-law and daughter when he officially retired, but he did plan to pass on the reins to his son-in-law once everything was settled.[1] The son-in-law had passed his master exams before the *Wende* with high honors, but he was forced to take the exams again with western examiners, again passing with high honors. His daughter continued to do the accounting, while still working part-time as a nurse in the Bernburg hospital, and one of his young grandsons announced that he didn't need to work hard in school, because he could always become a miller if all else failed.

By 1999, Mr. L. had made few changes other than purchasing a secondhand truck with which he could transport loose flour, nor did he or his son-in-law, who had officially taken over the business, have any plans for expanding flour production. The project of harnessing water power still had not progressed any further. But they were toying with the idea of selling ecologically pure products, like the old mill they visited in Bielefeld, in western Germany that had been transformed into a store for such products. One of the grandsons was planning to apprentice to become a baker and set up a bakery in an empty barn they owned.

As we have already mentioned, Mr. G., the western miller who owns the mill in our second example had a similar background to that of Mr. L. Although in our interview he deemphasized family tradition as a major motivating force in his career and even argued that such traditions could lead to a lack of innovation, his wife came from a long line of millers and his own father was a miller as well. His son was already working in the firm too and presumably would succeed the father. In addition, both millers were about the same age. What differentiated the two men perhaps more even than the fact that they lived under radically different regimes for forty years per se was first and foremost their access to resources, and second their expectations of the future of their enterprises.

While Mr. L. regarded himself as constrained by circumstances and willing to wait things out and moved only slowly to face new challenges, Mr. G. was constantly on the move and in search of new challenges. "In 1965, I entered into the parental business. At that time the turnover was one-twentieth of what it is at present. That was an interesting development too. But in the past five years the possibilities have become exhausted and so today the business is static. You can conduct business, of course, but it is more a task of administrating. My two brothers-in-law have experienced the same development during a period of restructuring in the western German milling industry, but that period has run its course. You can administrate, conserve, but you can't fashion. That is the decisive difference. . . . Where I come from everything has come to a standstill. You have your clients and if you try to win over new clients, you first have to take something away from someone else. The market doesn't expand. In fact, it may even be shrinking. Everything is terribly difficult, while here, this roaring development in the past year and a half has been an incredibly successful experience for someone who likes to do this kind of thing. It is fun." The partners began their search for new opportunities already in the first weeks after the border opened. "We looked around and slowly began to understand how the milling industry worked in the region. In the beginning, we were so shocked that we didn't have the courage to start something here. The first time I came here and saw how everything was in decay, the facades, in particular, I got scared. [Back home] we had our plants set up according to western standards, modern and visually attractive, or at least appealing, and we couldn't quite grasp what we were seeing here. But then we realized that the conditions here in this area were significantly better than at home, because of the large grain-producing area, the waterways and water power, as well as the closeness to large populations. We can reach at least 4 million inhabitants in a radius of 70 kilometers, all important factors for such an enterprise. In addition, whether we like it or not, the grain policies of the EU have been strongly influenced by the GATT negotiations. This means that in many regions of Germany, as well as in other EU countries, grain production will die out, because if grain prices are depressed it will no longer be economically viable."

So he purchased two mills from the *Treuhand* located some 50 kilometers from each other and began production in 1992. By then, the mill's market position had already deteriorated. As in many other industries, the bakers often found other sources of flour, often from the west, snubbing their erstwhile suppliers who had held monopoly power over them. Also, as Mr. G. explained, the mill managers had no idea how to deal with the new competitive situation. Indeed, the first persons they had fired were the employees of the marketing department. "They thought, 'What are marketing specialists good for?' Things work out on their own. It was very difficult, but we have been able to reassert ourselves in the market and now we have a good or even a very good market position." The firm immediately invested some 2 million DM in transportation vehicles and an additional 3 million to restore the power plant and repair the mill. Much of the latter investment was merely a stopgap measure to keep the plant running, for the plant was located in the middle of a small town, slowing traffic and creating noise. Nevertheless, they were able to increase production from 3,000 to 8,000 metric tons per month and were able to break even after only a year of operating at a loss. The plans were to build entirely new facilities in a more suitable location and begin producing there by the fall of 1995 for a total investment of 30 million deutsche marks. The mill purchased grain in a radius of some 30 kilometers and sold flour in a radius that averaged 70 kilometers to artisanal bakeries, large-scale baking factories, and some 20% to grocery stores who sold it to households. Like Mr. L., Mr. G. made contracts with the farmers in the region to grow specific varieties of grain. He was pushing farmers to grow rye. To this end he became a major promoter of the grower's association mentioned in chapter 8.

Mr. G. had few problems retraining workers. "I am sure that the workers had to get used to a different work organization and rhythm. But we had no problems whatsoever. People often say that the people here are less eager to work, but I can't confirm this. In no way. It is not the case." This didn't mean that they didn't have to reorient their practices. "For example, in the first days that I was here I saw that our best mechanic was sitting on a tractor and moving around trailers, a relatively simple task. When I asked him why he was assigned to this task he answered that it had always been like that and that he had a permit to drive a tractor and the others did not. I told him that from now on there was no such thing as a permit. Such a permit was only required to drive on public highways and that it made no sense for a mechanic to drive a tractor. He was too valuable for that. There were many things like that, but they did not really present a problem." The mill had a commitment to the *Treuhand* to employ forty workers. In 1994, it still employed sixty, but automation of the new plant was projected to entail the reduction of the workforce to forty.

The new mill, like the small eastern German one, was involved in plans to produce electricity too. The present power plant, on a major river, had to be totally rebuilt because the corrosive river waters had destroyed the turbine installed in the 1950s. It produced a little less than the mill's power consumption, but its capacity could easily be doubled, an option that the firm planned to undertake in the future. In addition, the enterprise purchased rights to several other locations where they planned to build plants if regulations permitted it. At the same time, the partners were involved in a large-scale bakery in Berlin. The firm also had plans to expand to Poland. They already had a German representative who sold flour there, taking advantage of the export subsidies and, in time, if the conditions were right, they thought that they might even produce flour there.

Mr. G. was one of the few entrepreneurs we revisited in 2001 who was highly satisfied with the way the enterprise had developed since the *Wende*. In spite of the fact that bread consumption in the east has stagnated or even decreased because of continued emigration to the west, they are producing six times as much flour as before, or 170,000 metric tons per year.[2] Mr. G. ascribes his good fortune to the fact that more and more industrial bakeries have opened branches along the major highways nearby rather than continuing to follow their earlier practice of supplying supermarkets from bakeries in the west. His partner alone has set up three such bakeries including one in Berlin which he can now supply with flour thanks to the brand new throughway connection. At the same time, he has been able to reopen small hydroelectric plants in two of the five sites he had purchased in addition to the one already in operation in 1993. It has not all been smooth sailing, however. Mr. G. deplores what he considers excessive bureaucratic procedures. For example, although the plans for the new mill had been fully approved by the authorities, they ultimately refused permission to open the plant, making a set of new conditions. Mr. G. decided to open the mill without permission for five years until the court decided the matter in his favor. Obtaining permission to rebuild the hydroelectric plants has been equally difficult. And he is still not sure that he will receive title to all the grain storage facilities he had purchased from the *Treuhand* but has not yet developed.

Mr. G. and his son's forays farther east have had mixed results. It was easy to sell flour in Poland but it was more difficult to collect the money, so they lost money. In contrast, they were able to export flour to Estonia where they have found more reliable customers. The younger Mr. G. became acquainted with them through a friend who moved to Estonia and built a small flour mill there. After working with the young importers for three years, they decided to purchase an existing flour mill together where they are producing rye flour. Their next project is to introduce better wheat varieties and to produce high quality flour. They plan to produce some 60,000 metric tons of flour and cover half of the country's needs. Mr. G. doubts that he will expand further into former eastbloc countries. Instead he plans to adapt his production more closely to the needs of the large-scale industrial bakeries who require special flour mixes to make buns that are just the right size to make standardized fish sandwiches. This will require close coordination with farmers who produce grain with a minimum amount of fertilizer and pesticides and further experimentation in the mill's laboratory and test-bakery. Mr. G., now 65, continues to make the weekly five hour commute between the west and the east, but is leaving much of the day-to-day operation of the enterprise to his son. His wife, who had managed the mills in the west has retired and they have rented the mill to others. The son, now married to another westerner, has moved permanently to the east. Except for one easterner who reestablished a firm in the region, most of their friendships continue to be with other westerners.

To conclude, the two millers, Mr. L. and Mr. G., were more aware of the fact that at least some sectors of the economy were again subject to centralized planning, albeit in new forms. Both firms stood to benefit from the promotion of clean forms of energy, which was, however, also subject to concerns about unduly altering river flows. Mr. G., whose mill was potentially large enough to export flour beyond the immediate region, was affected by the fact that grain prices were kept artificially homogeneous by the obligation of the state to purchase surplus grain. This made it uneconomical to ship grain from a region with a surplus to one with a deficit because the prices were the same regardless of supply, a fact that influenced Mr. G.'s decision not to enlarge his mill in the west and mill grain imported from the east instead of starting afresh in the east.[3] On the other hand, since the export of flour (but not of grain) to eastern European countries was subsidized by the state, it made sense for him to export flour to neighboring Poland. However, the most salient lesson taught by the two cases is the enormous potential for profit for a firm that has the necessary financial backing, and the bleak prospects of a firm with few financial resources, especially when combined with a shrinking regional market. For, while it is certainly true that the closure of many of the large state-run flour mills should have increased opportunities for a product that is sensitive to transportation costs, the fact that Mr. L.'s traditional clientele, the small-scale, artisanal, private bakers, were drastically reduced in number, diminished the importance of his social capital, the loyalty of his old clients. Perhaps too, the easterners' lack of familiarity with aggressive marketing strategies alluded to by Mr. G. was an impediment.

Thus the eastern miller was able to continue producing largely because, he continued to fulfill an interstitial role.[4] Nevertheless, as in the baking industry, a potential interstitial role in specialized niches remains for small-scale producers with modest aspirations.[5]

The Nature of Economic Networks

The network of linkages between farmers and their service providers and clients, a network that both shapes and is shaped by the identities of the various actors, has developed in peculiar ways. Before the *Wende,* firms, collectives, state enterprises, and the few remaining private firms alike were inserted into the wider economy in two ways: centralized allocation by the state and barter among firms. After the *Wende*, this kind of allocation was ostensibly eliminated, although, in actual fact, a different kind of allocation survived in the guise of maximum production quotas determined by the European Union and monopolies such as the local sugar factory. Otherwise, allocation in terms of input and output was replaced by new channels dominated by western firms that competed among each other but largely nipped emerging eastern competition in the bud. Barter for goods and services no longer played a role, but the social networks that were involved in such transactions became the basis for establishing a clientele. Whereas, in the past, these actors were interdependent because of scarcities resulting from the inefficiencies and the politics of the official system of distribution, today they are the basis for the flow of goods and services within a market system. In the process, the reciprocal links among all actors are in the process of being replaced by centralizing linkages to distributors and clients. The fact that linkages among firms that would be considered competitors in a market system (but did not regard themselves as such in the communist system) have not entirely disappeared is evident in the case of the electrician, who "borrows" a worker from a competitor when he can spare him. Of course, it is also very common in agriculture, where private farmers—but not cooperatives—engage in reciprocal exchanges of labor and equipment.[6]

The example of the electrician also shows another complex intertwining of GDR strategies of remedying the rigidities of a system of allocation with neo-liberal strategies for streamlining production. The electrician was rehired by the seed farm who had fired all its mechanics and electricians and outsourced such services, because he alone understood the intricacies of a nonstandard electrical system. In order to assist this former worker, they also provided him with cheap quarters for his enterprise. The seed farm thus continued to be tied to a specific individual to do the work for them.

In contrast, ties that would appear to be of a traditional nature because they involve economic links among individuals in the same locality actually represent innovative practices or harken back to the precommunist past. During communism, Mr. L., the small-scale miller, purchased all his grain from the state distributor. Direct arrangements with neighboring collectives would have been out of the question. Under the new order, he can make contracts with neighboring farms to produce to his specifications. In this, he is not alone. The western miller makes similar arrangements, and, in addition, he is an active organizer of the producer association. Interestingly, this role may not be a simple carry-over from *his* past, for it is unlikely to be widespread in the west. As some western farmers have pointed out to us, it presents a blatant conflict of interest, for producer associations should be geared toward generating rather than limiting competition among prospective buyers. It is consonant, though, with the paternalistic practices of the sugar mill that places its facilities at the disposal of the regional farmers' association.

But for the western newcomers, local strategies are just a small part of the creation of commercial networks. Rather, their orientation is toward linking the old and the new Germanies and to use their foothold in the new *Länder* as a jumping stone to enter the emerging or prospective markets farther east. Our eastern informants rarely made use of transnational linkages. An exception was the eastern German distributor of agricultural equipment cum manufacturer who was able to make use of the ties he had established while representing his firm in the FRG during the GDR period. We should note, however, that the GDR regime diligently tried to limit the construction of cross-national personal networks by moving him around at frequent intervals. In other sectors we have observed the importance of having relatives in the west, even though the GDR regime frowned on nurturing such ties. Surprisingly, though, we found few easterners who used personal ties with individuals in other former COMECON countries to promote their businesses. Eastern entrepreneurs seemed reluctant to engage in international travel other than to major trade fairs, a major part of their work routines.

Convergences and Divergences between Farmers and Their Associates

The inclusion of suppliers, service personnel, and clients in our analysis permits us to pinpoint the commonalities and differences between the various actors in eastern German agriculture and, more broadly, in the eastern German economy. It also brings to light the special position of eastern German agriculture in the region and beyond.

What differentiated farmers in collectives from employees of state farms and other state enterprises at the time of the *Wende* was their claim to the assets of the enterprise. A significant part of the land cultivated by collectives belonged to its members or to individuals who had moved away. The other assets belonged to the members jointly. This capital formed the basis for both the continuation of collectives in the form of modern cooperatives and the development of individual farms. The employees of state firms had no such base. And, since it was difficult to accumulate capital in the GDR that could be offered as collateral even in the form of real estate, would-be entrepreneurs from within such firms found it very difficult to convince lending organizations to underwrite their projects, even with the backing of the state. Not surprisingly, even highly skilled and motivated individuals sought westerners with capital to take over the firms, as did the *Treuhand*. Nevertheless, exceptional individuals, such as Mr. A., succeeded in obtaining the necessary financial support. Regretfully, these entrepreneurs have not always survived.

A second factor that explains the reluctance of many managers of state enterprises to strike out on their own was certainly also their inexperience in autonomous decision-making, exemplified in the difficulty Mr. S. professed in having the courage to make decisions on his own. To a much greater extent than LPG mangers, they were cogs in the wheels of large state bureaucracies. Although decision-making in agriculture was also top-down, and, as we have seen, often stupid and arbitrary, LPG managers had a long history of trying to maintain as much autonomy as possible. They also had to administer large budgets. Even some section heads of such specialty rubrics as growing hops were able to exercise a degree of independent control over their spheres of influence.

As a result of the foregoing, the pattern of western ownership and eastern management has been more prevalent outside agriculture. The western owners often attribute their success in the east to the combination of western capital and know-how with their eastern managers' social capital and the workers' resourcefulness and willingness to learn. In farming such arrangements are rare. They are restricted to the cultivation of land leased by westerners who arrive at the scene only during planting and harvest, the *Tiefladerbauern*, although some absentee landowners hire western managers, a practice that, according to one informant, may become more prevalent.

Since both managers of collectives and of state enterprises had to battle an ineffective system of allocation, they all shared a highly developed resourcefulness to make do with what was available and to deal with arbitrary political decisions. This did not mean, however, that they were not surprised and annoyed by the fact that the *Treuhand* and other agencies of the new state acted, in many ways, in a seemingly equally arbitrary manner. They had expected western institutions to be more transparent and democratic.

In both agricultural collectives and state enterprises individual economic activities during the GDR became the basis for post-*Wende* entrepreneurship. In the case of the agricultural collectives this took the form of keeping private livestock and in the case of state enterprises moonlighting and contracting for one's own needs.

In such sectors as the artisan trades, unlike agriculture (with extraordinarily rare exceptions), some private firms were allowed to continue in the GDR. Some artisans like the small miller were able to continue their activities without interruption. In contrast, the eastern individual farmers had to reach back to the precommunist past for a model of truly private farming. But the individual farmers were now able to enter a heavily subsidized sector of the economy, whereas the artisans had to relinquish a system of modest protection for their activities in the GDR and enter a heavily contested market arena. If they are not swamped by the homogenizing tendencies from the west, they may survive to serve the new local bourgeoisie that is again demanding specialized commodities and services.

More generally, the risk in many of the emerging enterprises outside farming was greater than in farming. Construction turned out to be a highly fickle sector that was characterized by a major boom in the early 1990s followed by a drastic downturn (see Buechler & Buechler 1999b). And the lack of capital led to a predilection for low capital activities that became commensurately crowded.

Last, but not least, the economic conditions in many sectors were characterized by the possibility of expansion and of achieving economies of scale. This opportunity was based, in part, on satisfying unfulfilled needs and, in part, on the fact that firms in the GDR that were continued or taken over were often gigantic. In agriculture, expansion has probably largely run its course. Farmers complain about the difficulty of obtaining access to additional land. Entrepreneurs in other fields are also encountering barriers to further expansion, but often have an eye out onto the possibility for expanding farther east.

CHAPTER TEN

Conclusion

When we first arrived in eastern Germany, we expected to find a blossoming of entrepreneurship made possible by the rapid replacement of a command economy by a capitalist one. We also expected to find that some of the social and economic forms developed in the GDR would not only survive the transition in some form but would assist the entrepreneurs in the adaptation to modern capitalism. While we did find such a blossoming, we also found a large degree of preemption of local developments by western firms with powerful lobbies, banks, and bureaucracies. Western German capitalism was less open to new actors in practice than it was in theory. As a result of the manner in which the east was taken over by the west we did not find many conversion experiences among easterners. We found, instead, that they weighed out loud the pros and cons of the transformation, following the models proposed by the west only to the extent they regarded them as beneficial or when they had no other choice. It soon became apparent to us that, whatever the consequences of the transformation, it would not result in a replica of either western German capitalism or, for that matter, global capitalism.

We realized that the past could not be erased overnight. But we did not expect the persistence of such communist institutions as agricultural cooperative production. We also did not expect the extent to which the new forms of agriculture would be inspired by *precommunist* models, some of which had not even survived in the west. So our task in this book was to illustrate the manner in which the multiple forms of agriculture, based on precommunist, communist and individualistic neo-liberal models, compete and intermingle.

Agriculture in eastern Germany developed both in consonance with and in reaction to political forces and actors. It is the product of the peculiar functioning of hegemonies. By their very nature, hegemonies do not function according to a unitary logic. Rather, in order to survive, they incorporate elements of competing systems with antithetical logics even if such an incorporation may ultimately contribute to their demise.

Counterintuitive twists result from the incoherent structure of hegemonies. In the GDR, state enterprises represented the dominant enterprise form. In them, the state was in complete control of production and distribution of surpluses. In contrast, agricultural collectives retained some of the legal characteristics of the cooperatives out of which they had evolved. Cooperativism similar to its present form was instituted in the late nineteenth century and became prevalent in agricultural production for a short period in the 1950s only to cede progressively to collectivization in the 1960s and 1970s. Forms of cooperativism survived in the sense that individuals continued to hold title to the land and that the collectives retained a degree of autonomy. Unlike in the Soviet kolkhoz, land in the GDR continued to be individually owned, even though ownership no longer had any practical significance for the owner. Also, the collectives retained some freedom of action not shared by state enterprises, in spite of restrictions imposed on the freedom of cooperatives, such as forced membership and later forced joining of collectives into larger units. The state condoned collectives but always retained an ambivalent stance toward them. At the same time, the state allowed private farming that emerged in reaction to the rigidities of the communist system, but kept it within strictly defined parameters, primarily by limiting access to land and other resources. Antihegemonic forces were thus tolerated and put to use to assure the survival of the system as long as their influence could be held in check. This the GDR regime was able to do much more effectively than other COMECON countries like Hungary. The GDR was characterized by a uniform ideal model that they could not fully implement.

After the *Wende*, the relationship between the farmers and the hegemonic forces became equally complex and ambiguous. Westerners did not uniformly support a single economic system. They applied several competing models of capitalism to eastern agriculture instead. The first model stressed the family as the basic unit of entrepreneurship and sought to replicate the small-scale family farm in the east. The second stressed economies of scale and corresponded more closely to corporate- or family-owned firms in other economic sectors that were characterized by greater reliance on outside labor.

Eastern independent farmers (along with private artisans) conformed to the first model. They were seen as representatives of the entrepreneurial spirit acclaimed by western German capitalism, and these farmers were given preference by the *Treuhand* for access to state land. While family farms became common, they did not become the dominant form. In spite of the subsidies and preferences offered, the number of eastern German farmers willing to try their hand at independent farming was lower than expected. Few easterners had ever run individual farms. Indeed, a background as a manager in a collective could be a more valuable asset than a background in individual farming. Of equal importance was the fact that few had sizable amounts of land, so access to credit was difficult. Nevertheless, all our eastern independent farmer informants shared a background in family farming that provided an incentive to take up an old tradition, as well as collateral for loans.

Eastern farmers claim that they are reestablishing precommunist traditions held in common with the westerners. Some aspects of this common tradition, such as the greater emphasis on family ties, are espoused without reservation. Thus, the eastern German independent farmers conformed most closely to the ideal model proposed by the west.

Western farmers who established farms in the east were closer to the second western model. With their greater access to large amounts of capital, they were able to engage in farming on a scale far larger than most of their counterparts in the west. While many of them continue to consider themselves family farmers, they are also participants in large-scale neo-liberal global capitalism.

The situation is further complicated by the attempted reintroduction of a model with feudal roots espoused by the returned nobility and western families with long-term eastern roots. They represent the familistic tradition archetypically, but also regard themselves as actors in a pan-German and global arena, with commercial ties extending to the German government, to multinational corporations, and to businesses overseas. Both emulated and viewed as anachronistic, they constitute influential lobbies in the west.

The collectives could not be accommodated within these western models. But the west was not able to reject them in their original form as cooperatives. Since agriculture was never fully nationalized in the GDR, this transformation was indeed plausible. One would have thought that, because of their large scale, collectives turned cooperatives would resonate with the western neo-liberal model. Also, the east and the west had a common history of cooperativism. In the west, cooperativism continues to manifest itself in *Raiffeisengenossenschaften* (purchasing and sales cooperatives first introduced by Raiffeisen in the nineteenth century) geared mainly to the purchase of inputs and to commercialization, even though these cooperatives have long ceased to create communities. This form of cooperativism has indeed been introduced in the east, although paradoxically other forms, such as hiring consultants cooperatively, fell by the wayside, ceding to more individualistic arrangements.

However, unlike consumer and distribution cooperatives, the rare production cooperatives in the west are often denigrated in the western literature as inefficient. Westerners considered production cooperatives in the east as a continuation of GDR collectivism, and therefore not worthy of support. Many easterners believe that the continuation of the debt incurred by agricultural collectives during communist times was a political move designed to bring about their dissolution. Indeed, few collectives in any other economic sector were transformed into cooperatives after the *Wende*.

The creation of a new hegemony that combines elements of the old order in the east with elements of the order in the west has been facilitated by the fact that agriculture in the European Union shares some of the features of a planned economy. In fact, the eastern German farmers are frequently reminded of the intrusive power of the state and supranational institutions even in the minute details of agricultural planning and compensation. Although cooperatives have been stymied by regional and national policies, those of the European Union have unwittingly favored them. The transformation is also taking place under the aegis of a welfare state that enabled firms to reduce their payrolls without creating poverty and social unrest. Centrally assured full employment of the GDR has been replaced by a centrally assured social safety net.

Our study reveals the extent of heterogeneity of both hegemonies and resistance to hegemonies. Resistance can be seen as based on flexible, often contradictory conceptualizations, and hegemony that currently exists in the former GDR is also ambiguous and indeterminate. Germany in the 1990s was an unsurpassed example of the contingent nature of both hegemonic power structures and, as we shall see, individual identities. Ambiguity and contradiction, far from constituting passing phenomena, are deeply rooted in eastern German agriculture and may well characterize its landscape for many years to come.

The new hegemony has global parameters, for the indeterminacies, ambiguities, and contradictions are, in part, related to the interplay among the local, national, and global levels. Some of the relationships are more direct, and others are more indirect, that is, mediated through the nation, in general, and western Germany, in particular, as well as through individual power brokers. The *Wende* provided all easterners and the farmers interviewed with the possibilities to travel and to communicate with persons beyond their borders. They took advantage of the offers by John Deere to travel to the United States and visit farmers there. They also began to pursue hobbies internationally, like training pigeons. But it is often in its indirect manifestations that globalization becomes most apparent. This is particularly true in state policies of crop subsidies and subsidies for fallowing and production ceilings that are directly influenced by European and global agreements. These manifestations of globalization impact regions differently. Powerful western agricultural lobbies induced the government to endow the eastern regions with lower production quotas than the west, thereby limiting the region's development potential. On the other hand, at least in the short term, certain EU policies also favored the eastern farmers who had access to cheap land.

The concerns that drove EU policies were mainly related to reducing and buffering global competition and making the EU more self-sufficient in terms of energy. Some, such as the promotion of wind-generated electricity, also were an answer to environmental concerns, although the farmers saw them as visually polluting the landscape. Finally, some measures were purely ecological. The Bernburg region, because of its high productivity, was one in which production was supposed to conform most closely to market principles. Only the most marginal land and land near rivers was managed predominantly according to environmental concerns. Ironically, it was the construction of a major throughway that led to the institution of the most far-reaching ecological measures in the area.

All the farmers and other entrepreneurs we interviewed supported the project of capitalism, but they varied in their assessment of the manner in which it had been introduced or imposed. None chose a model of agriculture that conformed to the type predicted by neo-liberalism. They all felt that differential access to the largesse of the state would determine their ultimate viability within the region, the new Germany, and the world.

All three types of farmers themselves have vested interests in the perpetuation of the new capitalist system since it has provided them with opportunities for economic and, in some instances, social advancement. By establishing their own farms, the individual eastern farmers have, at the very least, been able to maintain their own employment and often that of their spouses and members of the next generation as well. But they have done so in a local context that is characterized by women and men who are supported by retirement and early retirement, welfare, retraining, and workfare programs. Many of the unemployed and underemployed, although disappointed in their new roles, claim that they are better off financially in the new system and do not wish to turn back the clock, even if they speak with some nostalgia about the past. But they do observe with envy the development of the nouveaux riches in the east and the influx of western entrepreneurs, some of whom are wealthy and claim nobility. This differentiation has contributed to the perceived loss of community based on work and has yet to be replaced with community based on other criteria such as associational life or professional ties. For the cooperative managers it has meant the opportunity not only to maintain their previous positions but possibly to gain an economic standing that is superior relative to other members of the cooperative than the one they had enjoyed in the past. Finally, the western farmers found opportunities to start farming operations on a scale that they would never have been able to achieve in the west, or, in the case of the *Tiefladerbauern,* to add a second farming operation to their farms back home. Their orientation was more global than that of the eastern farmers, for they and, particularly, their western suppliers, had a broader view of the market.

At the time of our first and second field research periods, points of conflict and of contrast still dominated the interaction between the different types of farmers. The arena in which this self- and mutual definition unfolded was characterized by the confrontation between easterners who, while being the beneficiaries of generous development plans and welfare programs, were subject to the economic and political power of the westerners and westerners who were in a privileged position because the system they represented had led to far greater accumulation. The eastern farmers were now operating within a political and legal system essentially transposed from the west. Their conflict is manifested most strongly in the struggle for land and, more generally, between two philosophies of production, one predicated upon the extensive use of resources to maximize profits with little labor input, and the other on more labor-intensive and diversified strategies. These strategies are quite unanticipated. From the perspective of a Taylorist model of production, one would expect cooperatives, the successors of collectives that were based on achieving economies of scale by creating extremely large farms, to emphasize labor-saving strategies. However, as we have seen, enterprises under communism were inserted in an economy that stressed full employment and were geared toward hoarding inputs including labor to offset uncertainty and scarcities. After 1989, the collectives attracted members' votes for reorganization as cooperatives by stressing their continued commitment to retain jobs.[1] Conversely, one would expect farmers with a tradition of family farming to choose more intensive strategies commensurate with those employed by their families in the west.

Instead, they follow the Fordist model most closely. They have generally chosen to forfeit dairy farming and those crops that are particularly labor-intensive. They work with far fewer workers per hectare than their families did at home, doing most of the farm labor themselves, and they are highly mechanized. Indeed, some continue to live in the west and appear on the scene only when major farm work needs to be done. Theirs is a strategy commensurate with their position of power that has enabled them to take full advantage of government and European Union subsidies and financial backing, and the technical know-how to rent and cultivate much larger land surfaces than those of their families back home and those of the eastern independent farmers.

The actors in agriculture view their contribution to the local community and the region in different ways. The cooperatives regard themselves as upholders of local and regional concerns with employment, problematic as such a self-image may be in view of the dramatic reduction of rural employment. The eastern independent farmers regard their primary role as providers for specific local needs. Finally, the westerners see themselves as outsiders with a more global perspective, but also as permanent residents in the east attempting to broaden their local networks and contribute to local concerns. Some also regard themselves as returning patrons who are reestablishing (albeit self-consciously and uncomfortably) long-lost traditions of local leadership. While the present interaction among different farmers, particularly cooperative and individual farmers, is characterized by conflict, and their identities are formed in opposition to those of other actors, the identities also converge in many ways. Eastern German farmers have adopted the ideal of the market economy.[2] They present themselves as seasoned entrepreneurs, attuned to the exigencies of demand and profitability. They regard themselves as survivors in a brutal new system where, unlike others, they have been able to swim either because of their adaptability, or because they already regarded themselves as having learned to be flexible in communist times in order to bypass the communist system. This self-perception contrasts with the commonly held opinion that managerial skills are profoundly lacking in eastern Europe. Conversely, the western farmers adopted large-scale farming at first tentatively, but later more assertively. Large-scale farming was also characteristic of the precommunist past and certainly also emulates the strategy of former collectives/modern cooperatives. The eastern independent farmers are, perhaps unexpectedly, *most* likely to emulate the family farm of the west. Rather than copying their counterparts in the old *Länder*, they are drawing their inspiration both from precommunist agriculture and from counter-hegemonic individual farming during GDR times (integrated as it was into the socialist system). More generally, they reaffirm the central role of the family which, during the GDR, served as a haven from state encroachment. While corresponding to the designs of western planners, only a small minority of members of former collectives have chosen this path. The strategy of the diversified family farm is still more plan than reality.[3] At present, many employ the same extensive strategies as the westerners.

The ambiguities and contradictions within systems and the change from one to another is paralleled by the relations and sentiments of the actors involved. Depending on the context, we found our informants often held inconsistent views. They both deemphasized differences and accentuated them. Especially in the first years after the *Wende*, their interaction has been characterized by claims for compensation and by competition for land. Even in the latter part of the 1990s, cooperatives were continuing to lose land to other enterprise forms. However, in spite of these divisions, there has also been substantial cooperation among the different actors. Individual enterprises founded by former members of collectives often separated in an amicable atmosphere. In such instances, an individual farmer might rent buildings from the cooperative to which he once belonged and/or have the latter engage in specialized work for him. There are also instances of cooperatives and *Neueinrichter* from the west who do not regard one another primarily as competitors because the western farmer may, for example, have obtained most of the land from a privatized state farm or from a neighboring collective that was dissolved. Also, even when there is competition for land this does not necessarily preclude cooperation. Swapping parcels of land to prevent having to cultivate widely disseminated small surfaces occurs even among inveterate enemies. Similarly, as in the case of *Wiedereinrichter*, *Neueinrichter* from the west may ask cooperatives to carry out specialized tasks for them.

In addition to such exchanges, the lines between the types of actors are blurred because a variety of intermediate forms have emerged. For example, Luft (1997:64–66) describes a case where fourteen independent farmers formed six partnerships (GbR) and share space in a large barn taken over from a former collective, with each group owning their respective space. Following the pattern established before privatization, they engage in complementary aspects of livestock raising and help one another out in manifold ways. For instance, they consult with one another before making any purchases of machinery and the owners rent it out to the other groups at a preestablished rate. The six partnerships have also formed a sales and purchasing cooperative for the sale of quality piglets. This cooperative, managed by the dissolved collective out of which the units were formed, also does the separate accounting for each of the members. Further, the cooperative is the sole owner of a limited liability corporation (GmbH) that produces compost and cultivates the land of the GbR.

There are other ways in which the lines between the categories can become indistinct. For example, we encountered an instance where a cooperative manager ran a private farm on the side because the *Land* did not allow the cooperative to take over a demesne. In this instance, the cooperative furnished all the machinery for the manager's operation. In 1999, the differences between this farmer and other eastern farmers were further accentuated, but in new ways, when the manager was forced to quit his position in the cooperative but continued to have much of agricultural work done by the cooperative. His refusal to compete with the cooperative was also manifested in the fact that he did not petition the *Land* for the second demesne managed by the cooperative.

Such blurring of lines as well as the observation of the activities of one's enemies is leading to a more nuanced mutual assessment by the different actors. Let us return once more to our case studies and examine some of the statements made by our informants. They both corroborate the categorical "otherization" and negative pigeonholing that one would expect from the competitive struggle described in this book, but they also go beyond such stereotyping to a more dispassionate assessment of the competitive struggle and show an empathetic understanding of the difficulties faced by their competitors. Some even show mutual admiration.

The sentiment of lingering resentment and nostalgia, and grudging acknowledgment of a common future became clear in the words of the following informants. As we have seen in chapter 9, Mr. A., the eastern tractor dealer cum small-scale manufacturer, was both elated by the opportunities in the new era and critical of the brave new world. There were aspects of life in the GDR that he missed and might never be able to experience again. He compared the *Wende* with the conquest of Mexico by Cortés and, like many of our eastern German informants, argued that contrary to western views, western Germans profited handsomely from reunification. "People say that German unification has meant an enormous burden for the FRG. That is incorrect. One-third of Germany was put up for sale. Billions were transferred, whether in the form of real estate or whatever. Just look at the cashiered politicians [from the west] who have been able to reestablish themselves here. Even in the sciences. They have destroyed all research." He was fully aware that industries were often a haven for the unemployed. "But people did not perceive it in this manner. They were not aware of this fact." He missed the social institutions like kindergartens and the easy access to medical care, and—since he was very active in sports—the sports training and sports facilities. He also missed the relative economic equality. "The professor drove to work in his Trabant just like everybody else. The social gaps did not exist nor this damned consumer mentality."

Like some other entrepreneur informants he said that he had finally been forced into buying a new car at the urging of his workers, who presumably felt that the old car did not fit the image of a manager of a prosperous company. He felt that if capitalism didn't take care of the inequities and continued to "tread on the social aspects of life then some form of socialism will reemerge, hopefully not of the Maoist variety." In his work, he missed the human aspect too. "I have many business partners but I notice again and again, whether at meals or any other social occasion, that the human aspect is missing. If you inquire about the family or anything like that you are met with silence. They don't let you approach them in a personal manner. Today it's not because someone may belong to the secret service. It's merely an ingrained practice that managers have adopted. . . . They continue to talk business. Yes, there are time constraints but when it is eleven at night I go to bed and say, 'Let's continue tomorrow.' A measure of humaneness is simply missing. That separates us from them. We too will change. Our children will be like that. They are already growing up in a very different world. They have to deal with things in school that they had never known before. Both of [my teenage sons] take martial art lessons. What an idea! But I can't persuade them to drop it. It's even fun. Instead of taking up Goethe, they read about new types of martial arts and positions." Finally, he sees the need to strike a balance between a concern for his workers and to stand firm against any form of impropriety that could taint the name of his company and adversely affect it in an intensely competitive environment.

Mr. B., the western tractor dealer, holds a view of west-east relationships that underscores that of Mr. A. in our previous example. "In the media one constantly hears how the Wessis complain about the Ossis and vice versa and I notice that too among our people over there when I am back home over the weekend. In H. many complain about the east, and yet they have never been there. One has to get to know the people. One has to understand them. There is a different mentality. People here are humane and one has to go along with that. That is actually a positive thing. There is really no secret formula of what to do and what not to do. One has to apply a lot of humaneness and also honesty. Our interaction with the farmers over there was often characterized by an old German sport, whereby each side tried to best the other. Here a lot is possible but only by means of an honest relationship. Here something is occurring that has never happened to us before, namely, that a client does not begrudge us a thousand deutsche marks note. They tell us that they do not mind our profit as long as the service is good. In the west an old German farmer would begrudge us even a profit of 3 mark 50 on a combine. So things are different here, they are more honest." Unlike many westerners, Mr. B. is very positively inclined toward cooperative farming. He feels that the cooperatives are much better in managing large-scale farms than the western newcomers and decries the attempts of the authorities to destroy them. The relatively lower number of tractors and combines per hectare of land required in large-scale farming is offset by the fact that it is easier to make plans from year to year, so Mr. B. actually prefers to work with larger units.

The eastern farmers had similar contradictory views of the west. When asked whether he had western friends, one farm manager said, "Both very good ones and very bad ones. Out of a hundred, ninety crooks, five are on the path of becoming crooks, and with five you can have a good relationship. That's because here we have to deal with those whom they want to get rid of over there. The former managers of companies here suddenly found that they had nothing to say any longer. They were considered as belonging to cliques. I don't think that belonging to a clique is always a bad thing. . . . Such opinions [hurt]. [On the other hand] I have a very good working relationship with [the owner of a chemical factory], also personally. Such individuals did not come here to push others around, but we established contact with them through our work. They are good people. . . . [But then there are others] like the mayor here whom they called Dimido because he only was around on Tuesdays, Wednesdays, and Thursdays (*Dienstag, Mittwoch, Donnerstag*), but received the extra pay [for working in the East] plus 2,000 for lodgings. Hair raising! Reunification was handled all wrong. One has to get used to that."

More specifically, statements about *Neueinrichter* by cooperative managers often stress the fact that the westerners created few if any local jobs. Explaining why he wanted to find local people to take over demesne land that the *Land* insisted had to be cultivated by individuals and not by a cooperative, one cooperative manager said, "The cooperative would then be able to cultivate the land for that person. The farmers that come from the west rarely do anything with us. They do it all themselves or hire other Wessis who have already moved to the area permanently. They don't hire any of our people. We buy rape and wheat from one of them because he doesn't have any storage facilities, we are friendly with one another, but we would prefer it if someone from our own ranks would take over the demesne land."

The relationship between the eastern and western individual farmers is certainly colored first and foremost by competition for land. As an eastern *Wiedereinrichter* said, "They (the westerners) leased the land away from under us. I started out leasing land for 3 marks per *Bodenpunkt*. The Treuhand wants something over 4 marks; 3.5 would still work out, but 4 already makes things very tight. In Magdeburg [the westerners] offered as much as 7 to 8, because five farmers wanted it. They probably can't even live with that, but they absolutely wanted the land. Without the offers from the west we would have had it a lot easier. . . . The people here no longer shout hurray when a Wessi wants to rent something. There have been too many dirty business deals for that. People have sold land to persons who arrived with a Mercedes and a secretary and later they simply vanished [without paying for it]." He then described a case where a westerner had purchased a storage hall ostensibly in order to open a store for hardware and building materials. After forcing the community to put a new roof on the building he acquired it for 10,000 deutsche marks. Then he hired workers. But nothing further happened. The community wanted the building back and he finally agreed to sell it for 180,000 marks. Finally, to add insult to injury, he had arsonists burn the building down. The informant's daughter had seen some strange men at the site shortly before it went up in flames. Earlier that day the same men had asked for directions to get there.

Friendships are beginning to develop between eastern and western independent farmers, but, as indicated earlier, they continue to remark on the strange ways of their counterparts. Thus one eastern family that continues to live very frugally although they have a large and comfortable house, remarked on what they considered the profligate spending habits of their western farm neighbors, who had equipped the home they lived in with expensive amenities and furnishings.

The prevailing climate characterizing the relationship between the *Wiedereinrichter* and the cooperatives is captured by the following quote of an eastern German individual farmer: "We are not in direct competition with the LPG (sic). But there is a struggle for power going on because of the unresolved issues. As a result there is a lot of bad blood and there are attempts on the part of the LPG to incite hatred against us among the population."

On the other hand, *Wiedereinrichter* also recognize the difficult situation of the cooperatives. The same informant said, "They are beginning to fight for their land, but up till now the *Wiedereinrichter* have enjoyed priority with the *Treuhand*. . . . But the LPG (sic) finds itself in the same situation as I do. They need the land to survive and to secure credit. I need the income to pay my debts. The difference is that in the

LPGs a lot of people are involved and so their situation is in some ways more serious, even though they have reduced their workforce already. It does happen that some people don't talk to one another any longer because of the land. I can't complain myself. Everything went amicably. We exchanged land, measured and wrote contracts, but in other places it didn't work out that way. I am an honorary judge at the court in N. We recently dealt with a case where an LPG refused to give out the land. So we had to go to court and present our findings. We had to indict them to force them to release the land." Another *Wiedereinrichter* expressed resentment that the former LPGs were able to obtain free help from workfare programs. "We don't get any money, while they let the state pay for their people. I am not in agreement with that. They have a lot of land from before, and they don't give any of it up. I had 15 hectares of my own and had to work to get more. Now I have 135, but that is not enough and I can't get land from any other source. It's not easy. . . . I arranged with an owner to rent some land. . . . Right after the *Wende,* I had approached a man from the west; he is 75 years old. He is the only one of the persons who owns land here and lives in the west who wants to lease land to me. Most of them rent to westerners. But for three years now I have been trying to get [the cooperative] to release it. They refuse to give it up. Now it almost appears as though they will be able to keep it after all. We have one of those 'wonderful' land offices where everybody knows one another from before. I am small fry in all this and can't do anything against them. There is no justice."

Like that of the *Wiedereinrichter*, the westerners' view of the cooperative managers is, as we have already seen in chapter 8, colored by the particular circumstances of their competition for scarce resources, including land and sometimes buildings. Thus one nobleman describes his initial relationship to the cooperative that had tilled his family's ancestral land as one of patronage. "I said, 'I am willing to engage in any form of cooperation.' Then I began to devise all sorts of projects, indicating what they could do to safeguard jobs by establishing independent entities out of particular parts of the former collective. For example, I suggested transforming a cafeteria to serve meals on wheels. We discussed it. But nothing came of it because they didn't know how to establish business ties in the west. Then I tried to have [a western company] take over the workshop. That too failed. I also wanted to help them sell agricultural products. They produce peppermint tea here and I arranged for them to be able to sell peppermint tea [for a large sum of money] to a producer of tea bag tea. All this without asking for any compensation. I am just telling you all this so that you know that it is possible to work with me. But then they stopped this collaboration very abruptly by presenting falsified papers to the state council in order to buy my parental home. When someone snaps up your parental home in front of your nose, then all confidence vanishes." After the initial friendly welcome he had received from the collective manager, who had told him to come out of the cold into his home, the transformation had been jolting. "The things that are happening here are not worthy of the Federal Republic I knew from before. In western Germany they speak of dark Germany [*Dunkeldeutschland,* a play on the initials for *deutsche demokratische Republik*]."

Similarly, even in 1999, another western farmer spoke about the tense relationships with the cooperative managers arising from the competition for land. "In part, we have good contacts with LPGs (sic), but in part, the battle fronts are still drawn," he commented. "It's too bad because these LPG managers still have a lot of influence in the community and the old cliques are still there. That is troubling. It is amazing how well they have adapted. . . . (Also), here it is so that the LPG-farms, or rather their managers, form part of the *Bauernverband* (Farmer's Association) and the private farmers are part of the *Landvolkverband* (Rural People's Association). That is not a good development."

In contrast, another westerner and his wife who have not faced head-on competition with cooperatives expressed both understanding and admiration for the latter. "If I had established the farm I established here in the west, I would, per tradition, have hired at least two or three other workers. I know from my experience back home that in a small farm the ties between co-workers are so strong that when the pie becomes smaller, one doesn't simply fire workers, but out of moral responsibility one sets oneself the task to find something for them to do, even if it is not as lucrative.

"The cooperatives face an even worse dilemma. They have lost the land that we are cultivating and they had a relatively large labor force even before they lost that land. As a result of our arrival, they have been forced to concentrate their workforce on ever smaller amounts of land. So they face pressure from many sides. [The possibility of early retirement] has been a great boon for them. On the other hand, the workers over 50 were often their most productive ones, for they had the greatest sense of responsibility vis-à-vis the firm. [This said], the ties characteristic of a cooperative had receded more and more into the background and the collectives had become more of a corporation (*Kapitalgesellschaft*) where people could earn a living but where there was no longer the sense of equality they had felt in the beginning."

The couple admired the versatility of the neighboring cooperatives who were taking farsighted measures to counter the effects of further projected land loss. For example, they were building their own storage facilities and improving their already relatively high economic standing by marketing their produce directly, thereby adding value to their output. As Mrs. B. explained, "In the west everybody complains that the processing stage is taken out of their hands and that, as a result, the farmer earns only a small percentage of the value of the end-product, while here they simply say, 'In that case we will take [that stage] into our own hands.' And that is indeed what these LPGs are doing." Mr. B. felt that he would not be capable of emulating the cooperatives' example. "I notice that I am not suited for it and this is a much more serious problem than it would have been twenty years ago. I have miscalculated the amount of information required for this kind of marketing. One has to know the entire context and have the skills to carry it off, not just the education. That holds not only for me but also for the true *Wiedereinrichter*.

"In contrast, the cooperatives have learned [these skills] very rapidly, or at least that's what I hear from others." Mrs. B. added, "I believe that they also share ideas better and that they stick together." "We have three LPGs here that I feel are particularly strong," Mr. B. continued. "I would like to know more about the one in X. I heard that the manager was a former member of the agricultural council that became a persona non grata under Honecker and was therefore demoted to the position of leading a very backward collective. No one had expected him to succeed in making something out of it and yet he did and the farm still works well. One has to say that these people are born leaders who, regardless of the nature of the problems they face, manage to solve them after a relatively short learning period."

Similarly, another western informant empathized with the dilemma of the cooperative heads. "Of course there are points of contention between the western German farmers who have taken over land here and the cooperatives. The LPGs have administered the land that they were then forced to cede so that new farmers could get started. The LPG X that farmed 10,000 hectares and had to cede half of it is not happy about it. It is not easy for the cooperatives to lose more land every year. But, [eventually], a more sane relationship and a very good agricultural base will develop."

Some of the differences in the sentiments and views expressed by westerners are due to their heterogeneous backgrounds. The *Neueinrichter* from bourgeois families regard themselves as different from those of noble descent. The latter are well aware of the negative connotations of being a *Junker*, that is, a landed nobleman. As a descendant of an industrialist who invested in land and has some pretensions to elite status put it, "The *Junkertum* was a term of insult. People have made political capital out of that fact. During the CDU-plenary, Mr. Echternach said, 'Do you really want to give the *Junker* east of the Elbe their land back?' In fact the issue is really money. But one doesn't like to speak about that too much. To be sure, there were a few families who lived from *the proceeds from* agriculture rather than from their *engagement in* agriculture. They had their latifundia (large landed estates) but were not really involved in agriculture. They could live grandly because a few hundred years ago they had received the land from some king, whether they had deserved it or not. It was passed on from generation to generation and the regional populace was treated badly. If someone wanted to take that up again, he would be shipwrecked. But a majority of those families saw their role differently." Our informant then went into a long discussion about the positive social role the nobility played in the state of Mecklenburg and about how most of them did not feel that their name carried connotations of a privileged position any longer.

The manager for the two noblemen echoed this view of the positive role of nobility and felt that the local population, especially in the older generations, remembered the well-kept appearance of the estates owned or managed by the noblemen. But he was also aware of local sensibilities about the return of nobility. He was not too happy when his boss appeared with his friends in a horse-drawn carriage for a rural picnic. He said, laconically, "The perception of some people may have been somewhat negative." Similarly the nobleman who manages the farm that he has established himself told us in parting that he really would like to own and ride a horse but didn't dare to do so because he was sensitive to the image that would project. The same went for hunting.

The ambivalence toward the nobility is manifested not only in their relationship to the easterners but also in the way in which they are viewed by western farmers from nonelite backgrounds. As the informant who took over the demesne put it, "Even as a student, I was aware that different worlds came into collision. On the one hand, there were the sons of farmers who were studying agriculture because they had no prospects of inheriting the family farm, and then there were the old aristocratic families, where it was almost a tradition to study agriculture. That fact did play a role. They came from their parental estates, in part real *Junkers*, even in this present day and age. They joined exclusive fraternities, where their fathers and grandfathers had already been members. That does exist. Few of these [noblemen] come here themselves. Rather, they send managers, who do everything for them. Those who do come here are of course less exacting [in terms of their social and intellectual life]. One is, after all, deprived of one's intellectual circles." We did notice, however, that part of the appeal for *both* the aristocratic and bourgeois western farmers (and at least one former eastern manager of a cooperative) was the prospect of enjoying some of the accoutrements of an elite status, that is, living in a manor, being recognized as the successors of landed gentry, refurbishing a church, or sponsoring a feast.

We contend that it is these incipient, more nuanced understandings and the mutual emulation that may arise out of them that will play an important part in the dynamics of agriculture in the east and, in combination with the exigencies of an economic arena that is more susceptible to global, neo-liberal tendencies, will lead to developments that are different from those in the west for all the actors involved.

As in the analysis of earlier patterns of action that serve as models for emerging patterns, we must examine both the dominant trends in the self-definition and mutual definition of the farmers as well as the various countercurrents that contradict these trends but also enable them to survive by tempering their inadequacies. In both cases, the countertrends may influence subsequent developments out of proportion to their original importance. Just as repressed and marginal aspects of GDR economics became models for post-*Wende* developments, more nuanced understandings of other actors may become influential in the construction of interpersonal relations in the future.

NOTES

Chronology

Based on Darnton (1991:21–26) and Thiele (1998:25).

1. Introduction

1. We shall employ the term "cooperative" exclusively for enterprises following this voluntaristic model, reserving the term "collective" for enterprises conforming to the Soviet model of much lesser autonomy.

2. An example Arensberg gave of such sequences is the act of giving a gift A—>B followed by the act of reciprocating the gift B—>A, which describes simple systems of reciprocity (1972).

3. For example, in the introduction to an edited volume *Displacement, Diaspora, and Geographies of Identity*, Lavie and Swedenburg (1996) still employ the concept of "Eurocenter," connoting a unitary hegemony, to describe the clash between the hegemonic west on situations of contact with subaltern peoples.

4. For example, in an article on Mashpee Indian identity, James Clifford (1988) shows how a group of individuals can claim a unique (in this case ethnic) identity even when its members share few cultural characteristics that differentiate them from surrounding groups. He characterizes the Mashpee as "people who were sometimes separate and 'Indian,' sometimes assimilated and 'American;' their history was a series of cultural and political transactions, not all-or-nothing conversions or resistances" (1988:342; see also Gewertz & Errington 1991).

5. Similarly, Lavie's own discussion of the multifaceted and fractious nature of the culture of intellectual circles of Israeli Jews from Third World countries residing in Israel in the forementioned volume not only belies a unitary, hegemonic "Eurocenter," but points to the blurring of boundaries between the hegemonic and the subaltern. Her informants are as much representatives of

the hegemonic culture as they are representatives of Third World cultures. Indeed, many of the articles in Lavie and Swedenburg's volume can be interpreted in this manner.

6. We borrow from both Bourdieu's (e.g., 1972) concept of habitus and practice, and Sahlins' (1985) notion of practical reason. However, both these concepts underplay the importance of hybrid cultural forms and the creation of new cultural forms out of incipient behavioral regularities. In contrast, Arensberg's (1972) concept of emergence is more attuned to such possibilities.

7. For a similar definition, see Laclau and Mouffe (1982:105–106).

2. Identities

1. All names are fictitious.

2. Such pools had been established early in the push toward collectivization to serve adjacent small cooperatives and, secondarily, the then still remaining individual farmers.

3. As we shall see later, in the GDR, individuals retained title to land pooled in collectives but received only a nominal rent for it.

4. Mr. Rascher employed the term *überbetrieblich*, working "beyond the enterprise," a term he also used to characterize the pooling of equipment among neighboring farms, indicating that his base remained the family farm.

PART I. THE COMMUNIST PERIOD

3. The History of Collectivization in Agriculture

1. The latter figures, however, hide the fact that the *Land* of Mecklenburg had holdings of much larger size and Thuringia and Saxony much smaller holdings.

2. The first expropriations actually came during the Nazi period, when estates of dissidents were confiscated by the regime. This included the expropriation of the largest agricultural enterprise in Germany—the 10,000-hectare farm of the industrialist Carl Wentzel, who was executed for his involvement in the 1944 conspiracy against Hitler.

3. The new owners had to pay the equivalent of 1,000 to 1,500 kilograms of rye per hectare of land; 10% of this amount was to be paid by the end of 1945. The rest was to be amortized over a period of 30 years (Kramer et al. 1957:35).

4. This informant's indictment of GDR ideas about democracy did not mean that he believed that western-style democracy was any better, for he ended his diatribe with a swipe at modern German politics. "In my opinion, democracy is the biggest 'whore.' Just think about it. Staying with the case of Germany, no one has voted for the CSU party except Bavaria, so with what right does the CSU govern the whole of Germany? Because the party works together with the CDU it has even become the largest governing party. Democracy is a strange thing indeed!"

5. The difference between collectives and state enterprises came, in time, to be officially accepted as permanent in such communist contexts as Hungary of the late 1960s as manifested in such passages as the following quote of Fehér, cited in Lampland (1995:207–208): "Co-operative property is the joint property of the members—it is collective property. This property is more limited than public property but *it represents socialist property of the same rank*. So co-operative property is *socialist social property*, one of its forms of manifestation. . . . It naturally implies that in our economic policy the planned direction of economy must be founded upon the co-ordinated and simultaneous development of state and collective ownership representing two forms of social property."

4. The LPG

1. 40% of the livestock collectives had no specialization, while another 40% had more than one direction with one central specialty (Gollnick et al. 1990:27).

2. The GDR enjoyed a special status vis-à-vis the European Economic Community because the 1957 Rome Treaty allowed trade between the two Germanies to be considered "domestic trade" (Fulbrook 1992, 54).

3. Of course, at present, the difficulty of obtaining credit has the same limiting effect on production, particularly for indigenous producers. In the case of the former LPGs, these limitations are exacerbated by the official distrust and disadvantaging of production units regarded as survivals from the communist era.

4. It should be noted, however, that there were only seven farms that raised more than 6,000 pigs in the entire GDR (Schremmer 1990:43).

5. Ironically, the predicament of the animal production cooperatives may, on occasion, have been the result of their attempts to *overachieve* quotas in response to production incentives. The danger of responding to such incentives was not only that the promised premiums were often not paid at all, but that the new production levels achieved could simply become the new standard for future production. Such incentives could also lead to overproduction.

6. As a former member of an LPG put it: "The GDR was not allowed to develop anything. The Russian (sic) said 'This is how it is going to be.'" The GDR only manufactured 90 and 110 horsepower tractors, while those from the Ukraine had 150 to 220 horse power. Quality was less of an issue than availability and horse power.

7. In a planned economy, the need to keep a vehicle or a piece of equipment alive on paper was of considerable importance, for it established claims to parts, service, and ultimately replacement.

8. Hoarding was prevalent in all aspects of production. For example, herbicides were used more sparingly than planned and then secretly kept for the next year so that the quota would not be reduced.

9. One *Morgen* equals 0.6 to 0.9 hectares of land, depending on locality.

10. The Saale was highly polluted by uncontrolled industrial effluents. The pollution was so severe that during high winds foam would cover the neighbor-

hoods of the city of Bernburg lying along the river. Even in 1994, the river had a distinctive smell of detergents.

11. In some instances, traffic other than LPG vehicles was limited to bicycles. But even in those cases, roads that connected public facilities were open to the public.

12. State enterprises did not contribute to community projects in this direct manner.

13. However, they could not be carried forward beyond that year.

14. Conversely, an LPG was allowed to place money on a reserve fund and add it to the wage fund during years when income from its operations was low, a fact that contributed to wage security and stability. No limits appear to have been placed on the size of the reserve fund, for which the state bank paid up to 1.5% interest, but, since inputs were difficult to obtain and profit distribution was limited, access to the funds was difficult.

15. See also Luft 1997:15 fn 10, and Gollnick et al. 1990:72.

16. The Trabant, the basic GDR-manufactured car.

17. Individuals age 18 and older could petition for a garden area of 400 square meters. The productive use of such gardens was fostered by the state. Finally, the church cultivated much of its land by hiring individuals privately (Hohmann 1984).

18. He used the same expression as the one employed by Mrs. Feldmann, *antanzen*, with its connotation of dancing to the tune of someone else.

19. Another consultant stressed that some workers were allowed to take their vacations even during harvest time, enabling them to coordinate with those of their children who had school vacations at the time.

20. Other LPGs had similar arrangements. Friendships developed from these exchanges which, on occasion, have survived the *Wende*.

21. The relationship of workers to land cultivated of this firm was typical. Gollnick et al. (1990:39) figure an average of 325 to 400 workers for a 5,000-hectare collective and an additional 30% during the harvest.

PART II. POST-COMMUNIST PERIOD

Introduction: The Aftermath of Reunification

1. A Danish friend of Hans looked longingly at a practically new oil burning stove that had been cast on the street for disposal. Had he come to visit us by car rather than by train he would have taken it back to France with him for a weekend cottage he owns there.

2. The following discussion is based on two articles in the September and December 1990 issues of *Top Agrar DDR-Spezial,* by a western German lawyer, Hubert Feldhaus.

3. Later, the additional requirement was introduced that a majority of those members who owned land cultivated by the LPG also had to agree with decisions regarding the future of the LPG.

4. An additional reason why dissolution was usually a bad idea was the fact that old debts had to be deducted in full before the members could receive anything (Thiele 1998:162). In contrast, as we shall see later, cooperatives that were not dissolved may eventually escape paying most of these debts.

5. We did not encounter any farms operated by western corporations in our study area.

5. The Cooperative Solution

1. The original land restructuring law (*Landwirtschaftsanpassungsgesetz*) formulated by the Modrow government in 1990 only envisioned transformation of LPGs into cooperatives. Only the revised version of the law, instituted in 1991, allowed for other forms of incorporating and corporations (Laschewski 1998:53).

2. See, for example, Koch 1991.

3. In western Germany cooperatives of this type do not appear to be considered practicable. Rather, when farmers cooperate, they create common law partnerships (GbR). Although such partnerships allow the joint use of all means of production, they seem to be formed, in practice, by small farms between 30 and 100 hectares that cultivate between 150 and 400 hectares together (Riemann 1990:26–28).

4. These proportions are mandated by the Law on Agricultural Adaptation (see Feldhaus 1991).

5. LPGs sometimes eventually paid back excess inventory beyond the obligatory share or investment (*Pflichtbeitrag*) brought in by a new member of an LPG (Feldhaus, August 7, 1991).

6. Probably because he is in many ways still a convinced communist, he did not mention the fact that such sales would have made little sense in the GDR, since the land could not have been cultivated individually and there were severe restrictions on using agricultural land for a different purpose.

7. The set-aside program specifies that to be eligible for subsidies the land must have at least 18 land points (*Bodenpunkte*) on a scale of 100.

6. The New Cooperatives

1. Unfortunately all the enterprises established by the former director ultimately failed between 1995 and 1998 after a period of keeping afloat through layoffs and subsidies, a fact Mr. S. ascribes to the director's lack of training in commerce.

2. Some other informants also noted a renewed demand for potatoes.

3. For another example, see Hartmann 1995:100.

4. Unlike individual farmers, all cooperatives, including farming cooperatives, are subject to a tax on trade. However, the tax imposition on farming cooperatives was postponed until at least 1995 (Hartmann 1995:65).

5. Some 26,000 deutsche marks.

6. The owner of the land receives an additional annual rent of 4,500 DM. However, the cooperative owns the land under only one of the windmills.

7. They were also disadvantaged in terms of taxes compared to individual enterprises (see Hartmann 1995:137).

8. For a detailed description of these loans, see Thiele 1998:198–201.

9. According to a survey carried out by a private firm in 1993, 88% of the 400 large (over 300-hectare) farms surveyed planned to undertake investments during that year (85% in Sachsen-Anhalt), even though few felt that the investment climate was favorable. Almost 63% of these firms planned to invest over 150,000 DM (Top Agrar Spezial, April 3, 1993, 8–10).

10. Economies of scale permitted cooperatives analyzed by Hartmann (1995:162) to obtain price advantages of up to 20% for machinery and 10% for other inputs compared to usual rates.

11. Similar observations were made by another cooperative manager, who says that he uses John Deere tractors in and out of his cattle barns every single day of the year with few repairs. He also praised the machines' tolerance for inexperienced drivers. Such considerations are of particular importance in large farms where equipment is under constant use. He gave a similar assessment of Russian tractors. He considered the cooperative to have been particularly fortunate to have been able to purchase a 220 horsepower tractor for only 30,000 marks just after the *Wende*. "Now, one has to pay 800 to 1,000 marks per horse power, while we paid only 150. We will continue to buy and use these machines."

12. See, for example, Top Agrar Spezial, November 25, 1991, 3.

13. In 1999, this cooperative also ceased producing hops because the area where it grew them, with particularly favorable microclimatic conditions, was destined to become part of a natural park.

14. An herb used as a condiment, for its medicinal properties, and as a bath salt.

15. Black bacteria with medicinal uses that grow on grain.

16. Some of the cooperatives reproduced the herd themselves, others purchased young cows.

17. The fact that the accord between the German government and the Soviet Union stipulated the purchase by the latter country of an amount of beef that roughly corresponded to the number of cows by which the eastern German herds had to be reduced to adjust them to the new quota system, kept the price the cooperatives received at acceptable levels (Neue Landwirtschaft, September 1990).

18. In the following year, the quota was subjected to a further 2% reduction (Top Agrar Spezial, December 23, 1991, 3).

19. The longer fattening period also makes the meat more dense.

20. According to official statistics, 345 (or 67%) out of the 513 workers listed as unemployed in agriculture in the *Kreis* of Bernburg in June 1993 were women (Arbeitsamt Dessau). While the unemployed men in agriculture represented 7.2% of the total unemployed men, the corresponding figure for women was 9%.

21. If one includes the average two or three drivers who are permanently or long-term (over 6 weeks) disabled, the ratio, between 8.33 and 8.88 drivers per 1,000 hectares, would still be substantially lower than before.

22. Interestingly, in one cooperative two men seemed to share the position of executive manager. Both were present at the interview and when sales representatives came to visit.

23. It should be noted, however, that managers were more likely to participate directly in field work before the final consolidation of LPGs in the mid-1970s.

24. The easterners' emphasis on the collective rather than on the individual is vividly illustrated by Berdahl's (1999:184) description of a community project to introduce villagers to silk-painting. Instead of engaging solely in their individual projects, the "women moved from project to project, picking up a paintbrush or glitter and adding whatever they saw fit to each other's artwork."

25. For obvious reasons, former managers of collectives did not mention receiving special consideration because of their position, but a member of a cooperative who had become independent accused the manager of his collective of having his private home constructed by the collective's construction brigade, presumably at the collective's expense.

7. The Eastern Independent Farmers

1. Schmidt (1996) quoted in Laschewski (1998:35) confirms our observations regarding the importance of individual farming during the GDR as a basis for modern individual farming ventures, because elements of typical peasant farming survived in these individual operations.

2. As we shall see later, some farms, particularly those operated by westerners, are much larger.

3. For an extended discussion of the strategies of the cooperative managers to minimize payments to their members, see Thiele 1998, chapter 4.

4. A number of German software companies have developed special programs to facilitate keeping track of complex land rental arrangements and to integrate them with other aspects of agriculture (Brosthaus 1991).

5. Sometimes farmers actually lost land because in GDR times their land had been used for a quarry or a highway for which they had never received adequate compensation. Such inequities have not been redressed after the *Wende*.

6. Since the figures are not broken down by origin of the farmer, it is impossible to judge whether they hide a bimodal distribution. It would seem to us that the share of the eastern farmers' own capital would be substantially lower than the average given here.

7. The *Land* of Sachsen took steps to solve this problem by forcing the local banks to provide a written agreement to give loans already approved by the subsidizing authorities. In addition, it provided debt-consolidating loans to farmers who were in temporary financial trouble through no fault of their own (Grund 1993).

8. We made similar observations in families that had established other types of enterprises: factory owners, building contractors, artisans.

9. In other words, he feels that it would not be economically viable.

10. Similarly, Berdahl (1999:43) found that, by 1992, in the rural border community she studied, few individuals still raised their own pigs.

8. The Western Farmers

1. The corresponding figures for eastern Germany as a whole are 7% and 5%, respectively.

2. This option did not apply to expropriated land that had been parceled out to refugees and small-scale local farmers in the land reform of 1949.

3. When we interviewed this family two years later, they had torn down the old house and were already living in a simple, modern, nicely furnished suburban-type house.

4. His father owns 67 hectares of land and rents another 25.

5. Without this farm, the average would be 520 hectares compared to 252 for the easterners.

6. This is somewhat of an exaggeration, for a hectare of land costs more like 9,000 to 14,000 deutsche marks.

7. One farm was a father-son partnership. In another case, the farmer was in constant telephone contact with his father and brother, both of whom were farmers in the west. The two brothers also shared an apprentice and the father helped during peak times in the seasonal cycle.

8. Of the 7 western farmers, 2 worked with only 1 worker, 2 worked with 2, 1 worked with 2 and an apprentice, the contract farmer worked with 9, and the farm manager with 10 workers, if one includes all the subsidiary activities. Of these, only one came from western Germany. In addition, the farm manager hired a western student on a temporary basis.

9. Although the partner of the Dutch farmer specialized in dairy farming.

10. The literature on western German and precommunist eastern German rural communities stresses the importance of associations in the lives of farmers (see, for example, Catt 1986:152; Wilke 1986:178–179).

9. The Service/Industrial Context of Agriculture

1. The traditional nature of the family enterprise is also manifested in the fact that the family was also involved in agriculture. Later, it continued to raise pigs and ducks. Today, the family still raises its own pigs and chickens.

2. Turnover has not increased as much because of the price policies of the EU that reduced the price of flour by half.

3. However, it also does not make sense to transport flour over long distances, because, as we have already seen, flour, unlike grain, is generally shipped

in specialized vehicles, making it impossible to use the same trucks to transport other goods on the return trip.

4. As we have shown in our analysis of the bakers of Bernburg (Buechler & Buechler 1999a), artisanal or semiartisanal producers were sometimes allowed to play independent, but strongly circumscribed roles. Ironically, the *Wende* hastened their demise, in spite of an ideology that favored entrepreneurship at all levels. Like most bakers, most private millers have shut down their operations, making way for large enterprises.

5. The bakers held such a niche because it is difficult for large-scale bakeries to meet the quality standards and standards of service possible on a smaller scale to fill a culturally specific need: the provision of fresh bread in easily accessible sales facilities.

6. We do not claim, of course, that such exchanges are characteristic of eastern farmers alone. Farmers in the west commonly engage in such practices too.

10. Conclusion

1. See also Koester and Brooks 1997:20. In this emphasis, they are similar to the eastern managers of state firms in other sectors of the economy who have made considerable efforts in finding employment for their former workers. In those cases, however, the firms usually did not have the option to achieve this by retaining or expanding into subsidiary activities. Instead, they helped their employees to establish independent firms.

2. The "German model" of capitalism has entailed a partnership between management and strong labor unions in both the private and the public sectors that assure workers their basic social needs, including adequate income and health care, and comprehensive, free vocational training. It also guarantees them a voice by means of organized representation in the workplace on supervisory boards and in work councils (Turner 1997:3–5).

3. It should be noted that farms in the west also tend to be specialized in either crop farming or animal husbandry.

GLOSSARY

Absahner der Nation	persons who cream wealth off the top of the nation
Agrar-Industrie-Vereinigung (GDR)	agro-industrial complex
Bauernverband	Farmers' Association: one of the two pan-German associations
bei uns	(lit. "here"): in our time and place (in the GDR)
Bezirk	district
Bodenanteil (GDR)	land rent
Bodenpunkt	point on a soil evaluation scale, with 100 points being equivalent to the best land in the nineteenth-century German *Reich*
Bodenreform	Soviet-instituted land reform of 1945
Boden-Verwertungs-und-Verwaltungs-Gesellschaft [BVVG]	agency in charge of privatizing land (successor to *Treuhand*)
Deputatenmilch	milk produced by landed estates for their workers
Domäne (Staatsdomäne)	demesne
drüben	over in the west
Dunkeldeutschland	"dark Germany": derogatory western term for the GDR (play on words on the initials DDR)
eingeschriebene Genossenschaft [eG]	legally constituted cooperative
Erzeugergemeinschaft	producers' sales cooperative
Flurbereinigung	land consolidation program

freie Spitzen (early GDR)	freely disposable production (after obligatory quota was fulfilled)
Gesellschaft bürgerlichen Rechts [GbR]	common law partnership
Gesellschaft mit beschränkter Haftung [GmbH]	limited enterprise
Gnadenschiedskomission (GDR)	pardon commission
Hausschlachte	home butchery: sausage made in artisanal production
Heimat	Fatherland
Hof	(lit. compound): farm, house, and barns
Junker	nobleman
Junkertum	nobility
Konsum, Konsumgenossenschaft (GDR)	state-owned grocery chain
Konzept	feasibility plan
Kooperation (GDR)	entity in charge of coordinating crop and livestock cooperatives
Kreis	county
Land (pl. Länder)	federal state
landwirtschaftliche Produktions-genossenschaft [LPG] (GDR)	agricultural collective
Landwirtschaftsan-passungsgesetz	Law of Agricultural Adjustment promulgated in 1991
Landvolkverband	one of two pan-German farmers' associations
Lehr-und Wanderjahre	time of study and travel of a student or apprentice
LPG Typ I (early GDR)	collective in which only agricultural land was held jointly
LPG Typ II (early GDR)	collective in which land and machinery were held jointly
LPG Typ III (GDR)	collective in which all assets were held jointly
Maschinen-Traktoren-Stationen [MTS] (GDR)	farm machinery pools

mittelständisch	bourgeois
Morgen	measure of land corresponding to 0.6 to 0.9 hectares, depending on locality
Nachfolgebetrieb der LPG	successor enterprise to a collective
Neueinrichter	newly established farmer in the east (usually westerner who did not have land in the area)
Ossidenken	east German mind-set
Pflichtbetrag (early GDR)	production quota to be delivered to the state
Raiffeisen-Genossenschaft	cooperative of the type introduced by Raiffeisen in the mid-nineteenth century
Rat des Kreises	county council
Ratsvorsitzende	council president
Revisionskomission	board of trustees
Stammtisch	regular gathering of a group of individuals at a pub
Rotlichtbestrahlung (GDR)	"exposure to red (infra-red) light": day of heavy communist propaganda at an official retreat
Tiefladerbauer	"flatbed truck farmer" who cultivates land mainly in absentia, appearing only for major agricultural tasks with his equipment loaded on a flatbed truck and hiring resident labor in between
Treuhand(anstalt)	agency in charge of privatization; also used as synonym for BVVG
unsere Zeit (syn.: bei uns)	our times (during the GDR)
Verein	voluntary association
Vereinigung der gegenseitigen Bauernhilfe [VdgB] (early GDR)	farmers' mutual aid association instituted in 1946
volkeigener Betrieb [VeB] (GDR)	state-owned enterprise
Vollbeschätigte Einheit [VBE]	unit of full-time work
Vorstand	executive committee
Wiedereinrichter	"reestablisher": person with land in the east who reestablished a farm there
Wiedereinrichtungsplan	plan presented to the government to reestablish a farm (required to obtain subsidies and credit)

BIBLIOGRAPHY

Agrarbericht

1999 "Agrarbericht 1999." *Neue Landwirtschaft*. March, 19.

AID-Informationen für die Agrarberatung, Bonn

1994 *Auswertungs-und Informationsdienst für Ernährung, Landwirtschaft und Forsten*, 2(7), 3, 5, 6.

Altmann, Torsten

1993 "Milchqualität: Vielen Betrieben droht die Liefersperre." *Top Agrar Spezial*, December 18, 4–9.

Arensberg, C.

1972 "Culture as Behavior: Structure and Emergence." *Annual Review of Anthropology*, *1*, 1–26.

1981 "Cultural Holism through Interactional Systems." Presidential address for 1980. *American Anthropologist*, *83*, 562–581.

Berdahl, Daphne

1999 *Where the World Ended: Re-Unification and Identity in the German Borderland*. Berkeley: University of California Press.

Berndt, Roswitha

1994 "Das Territorium Sachsen-Anhalt in der Weimarer Republik." In *Geschichte Sachsen-Anhalts III: Bismarckreich bis Gründung der Bezirke 1952*, pp. 81–134. Landesheimatbund Sachsen-Anhalt e.V. Munich/Berlin: Koehler und Amlang.

Berthold, Theodor

1972 *Die Agrarpreispolitik der DDR: Ziele, Mittel, Wirkung*. Osteuropastudien der Hochschulen des Landes Hessen, Series I, Vol. 53. Zentrum für kontinentale Agrar-und Wirtschaftsforschung der Justus Liebig-Universität. Berlin: Duncker & Humblot.

Bornemann, John

1991 *After the Wall: East Meets West in the New Berlin*. Harper Collins, Basic Books.

Bourdieu, Pierre

1972 *Outline of a Theory of Practice*. Cambridge: Cambridge University Press.

Brosthaus, Gerburgis
1991 EDV-Programme für Acker, Stall und Büro." *Top Agrar Spezial*, December 23, 8–10.

Buechler, Hans, & Judith-Maria Buechler
1995 "The Many Faces of Agricultural Privatization in Eastern Germany." *Anthropology of Work Review*, *14*(3&4), 32–39.
1999a "The Bakers of Bernburg and the Logics of Socialism and Capitalism." *American Ethnologist* *26*(4), 799–821.
1999b "Constructing the East: The Building Trades in Eastern Germany." Paper presented at the 98th Annual Meeting of the American Anthropological Association, Chicago, November 17–21.
2000 "Farm Managers and Social Identity in Sachsen-Anhalt." *Identities*, *7*(1), 39–83.

Bundesministerium für Ernährung, Landwirtschaft und Forsten
1990 "Investitionsförderung: Für bäuerliche Familienbetriebe und zur Umstrukturierung von LPGen."

Catt, C.
1986 "Farmers and Factory Workers: Rural Society in Imperial Germany: The Example of Maudach." In *The German Peasantry: Conflict and Community in Rural Society from the Eighteenth to the Twentieth Centuries*, ed. R. J. Evans & W. R. Lee, pp. 129–157. London: Croom Helm.

Clifford, J.
1988 "Identity in Mashpee." Chapter 12. In *The Predicament of Culture*, pp. 277–346. Cambridge, MA: Harvard University Press.

Comaroff, Jean, & John Comaroff
1986 "Christianity and Colonialism in South Africa." *American Ethnologist*, *13*, 1–22.
1991 *Of Revelation and Revolution: Christianity, Colonialism, and Consciousness in South Africa*. Vol. 1. Chicago: University of Chicago Press.

Creed, Gerald
1998 *Domesticating Revolution: From Socialist Reform to Ambivalent Transition in a Bulgarian Village*. University Park: University of Pennsylvania Press.

Darnton, Robert
1991 *Berlin Journal 1989–1990*. New York: Norton.

Dennis, M.
1998 "Family Policy and Family Function in the German Democratic Republic." In *Social Transformation and the Family in Post-Communist Germany*, ed. Eva Kolinsky, pp. 37–56. New York: St. Martin's.

De Soto, Hermine, & Christel Panzig
1995 "Decollectivization to Poverty and Beyond: Women in Rural East Germany Before and After Unification." In *East European Communities: The Struggle for Balance in Turbulent Times*, ed. David Kideckel. Boulder: Westview.

Dodds, Dinah, & Pam Allen-Thompson, eds.
1994 *The Wall in My Backyard: East German Women in Transition.* Amherst: University of Massachusetts Press.

Feckl, Ute
1997 *Untersuchungen zur Einkommens-und Finanzlage von Agrargenossenschaften-Stand, Zusammenhänge und Ausblick.* Gießen: Schriften aus dem Institut für ländliches Genossenschaftswesen an der Justus-Liebig-Universität Gießen.

Feldhaus, Hubert
1990 "Genossenschaftsbauern haben jetzt mehr Rechte." *Top Agrar DDR-Spezial*, September 21, 24–25; May 10, 18–21.
1990 "Was LPG-Mitgliedern jetzt zusteht." *Top Agrar DDR-Spezial*, September 21, 24–25.
1991 "Welche Ansprüche haben ausscheidende Mitglieder jetzt an die LPG?" *Top Agrar Spezial*, August 7, 6–13.
1991 "Wie das Eigenkapital bei Abfindungen ermittelt wird." *Top Agrar DDR-Spezial*, August 7, 14–21.

Fink, M., & P. Uphoff
1994 *Auswirkungen der Wirtschafts-und Währungsunion auf die Beschäftigungssituation, Lebenshaltung und Versorgungslage ländlicher Haushalte in den neuen Bundesländern. Forschungsbericht 1994. Institut für Strukturforschung.* www.fal.de/german/institutes/research/ber1994.htm

Fuchs, Klaus
1994 *Reorienting the Cooperative Structure in Selected Eastern European Countries: Case-study on the Former German Democratic Republic.* Central and Eastern Europe, Agriculture in Transition 3. Rome: Food and Agriculture Organization of the United Nations.

Fulbrook, Mary
1992 *The Two Germanies, 1945–1990: Problems of Interpretation.* Hong Kong: Macmillan.

Gewertz, D., & F. Errington
1991 *Twisted Histories, Altered Contexts: Representing the Chambri in a World System.* Cambridge: Cambridge University Press.

Gollnik, Heinz, Peter Wissing, & Jürgen Heinrich,
1990 *Die Landwirtschaft der DDR Ende der achtziger Jahre—Bestandesaufnahme und Übergang zur Marktwirtschaft. Agrarwirtschaft*, Sonderheft 126.

Grabher, Gernot
1992 "Eastern Conquista: 'The Truncated Industrialization' of East European Regions by Large West European Corporations." In *Regional Development: A Contemporary Response*, ed. H. Ernste & V. Meier. London: Bellhaven.

Gramsci, Antonio
1971 "Selections from the Prison Notebooks of Antonio Gramsci," ed. and trans. Quintin Hoare & Geoffrey Nowell Smith. New York: International Publishers.

Grund, Manfred
1990 "Aus für Thüringens Schweine-Großstall." *Top Agrar DDR-Spezial*, December 7, 5–7.
1991 "Schafhaltung Licht am Ende des Tunnels?" *Top Agrar Spezial*, January 25, 6–10.
1993 "Fünf vor Zwölf für Bauer Krumm." *Top Agrar Spezial*, January 16, 6–8.
1994 "Fett oder Fleisch—Schweinezüchter am Scheideweg?" *Top Agrar Spezial*, January 20, 4–6.
Gulliver, P.
1971 *Neighbors and Networks*. Berkeley: University of California Press.
The Hajj, S. Lavie, & F. Rouse
1994 "Notes on the Fantastic Journey of the Hajj, his Anthropologist, and her American Passport." *American Ethnologist*, *20*, 363–384.
Hartmann, Bärbel
1995 *Wettbewerbsfähigkeit landwirtschaftlicher Produktivgenossenschaften in den neuen Bundesländern. Dokumentiert anhand von Fallbeisbeispielen*. Gießen: Schriften aus dem Institut für ländliches Genossenschaftswesen an der Justus-Liebig-Universität Gießen, Heft 39.
Henning, Friederich-Wilhelm
1988 (1978) *Landwirtschaft und ländliche Gesellschaft in Deutschland, Band 2 1750 bis 1986*. 2., ergänzte Auflage. Paderborn: Ferdinand Schöningh.
Hoell, Günter
1990 "Ziele der Unternehmen neu bestimmen: Was erwartet die DDR-Landwirtschaft auf dem Weg in die EG?" *Neue Landwirtschaft*, *5*, 5–6.
Hohmann, Karl
1984 *Entwicklung und Bedeututng der privaten Agrarproduction in der DDR*. Berlin: Fs Analysen.
Horz, Gerhard
1961 *Die Kollektivierung der Landwirtschaft in der sowjetischen Besatzungszone Deutschlands unter besonderer Berücksichtigung des sowjetischen Vorbildes: Ziele—Methoden—Wirkungen*. Dissertation in Economics and Social Science, Freie Universität, Berlin.
Hübner, Hans
1994 "Sachsen-Anhalt 1871–1900." In *Geschichte Sachsen-Anhalts III: Bismarckreich bis Gründung der Bezirke 1952*, pp. 41–80. Landesheimatbund Sachsen-Anhalt e.V. Munich/Berlin: Koehler und Amlang.
Humphrey, Caroline
1983 *Karl Marx Collective: Economy, Society and Religion in a Siberian Collective Farm*. Cambridge: Cambridge University Press.
1998 *Marx Went Away—But Karl Stayed Behind*. Ann Arbor: University of Michigan Press.
Kideckel, D.
1993 *The Solitude of Collectivism: Romanian Villagers to the Revolution and Beyond*. Ithaca: Cornell University Press.

Koch, Walter

1991 *Der Genossenschaftsgedanke F.W. Raiffeisens*. Würzburg: Creator.

Koester, Ulrich, & Karen Brooks.

1997 *Agriculture and German Reunification*. World Bank Discussion Paper No. 355. Washington, DC: The World Bank.

Kolinsky, Eva, ed.

1998 *Social Transformation and the Family in Post-Communist Germany*. New York: St. Martin's.

Kornai, J.

1992 *The Socialist System: The Political Economy of Socialism*. Princeton: Princeton University Press.

Kowalewsky, H.-H.

1990 "Wie können wir unser Gülle-Problem lösen?" *Top Agrar DDR-Spezial*, August 6, pp. 6–9.

Kramer, Matthias et al.

1957 *Die Landwirtschaft in der Sowjetischen Besatzungszone: Die Entwicklung in den Jahren 1945–1955*. Bonner Berichte aus Mittel- und Ostdeutschland. Bonn: Bundesministerium für gesamtdeutsche Fragen.

Laclau, Ernesto, & Chantal Mouffe

1982 "Recasting Marxism: Hegemony and New Political Movements: Interview with Ernesto Laclau and Chantal Mouffe." *Socialist Review, Number 66, 12*(6), 91–113.

1985 *Hegemony and Socialist Strategy: Towards a Radical Democratic Politics*. London: Verso.

Lampland, M.

1995 *The Object of Labor: Commodification in Socialist Hungary*. Chicago: University of Chicago Press.

Laschewski, L.

1998 *Von der LPG zur Agrargenossenschaft: Untersuchungen zur Transformation gemossenschftlich organisierter Agrarunternehmen in Ostdeutschland*. Berlin: Sigma (Rainer Bohn Verlag).

Lavie, Smadar

1996 "Blowups in the Borderzones: Third World Israeli Authors' Gropings for Home." In *Displacement, Diaspora, and Geographies of Identity*, ed. S. Lavie & T. Swedenburg, pp. 55–96. Durham: Duke University Press.

Lavie, Smadar, & T. Swedenburg

1996 "Introduction: Displacement, Diaspora, and Geographies of Identity." In *Displacement, Diaspora, and Geographies of Identity*, ed. S. Lavie & T. Swedenburg. Durham: Duke University Press.

Ludz, Peter Christian, & Johannes Kuppe

1979 *DDR Handbuch*. 2nd ed. Bundesministerium für innerdeutsche Beziehungen. Köln: Verlag Wissenschaft und Politik.

Luft, Hans

1997 *Landwirtschaft Ost kontra Treuhandmodell*. Berlin: Dietz.

1998 *Blickpunkt Landwirtschaft: Zum Transformationsprozeß osdeutscher Agrarstrukturen*. Frankfurt: Peter Lang.

McDonald, Maryon
1993 "The Construction of Difference: An Anthropological Approach to Stereotypes." In *Inside European Identities: Ethnography in Western Europe*, ed. S. Macdonald, pp. 219–236. Providence: Berg.

Mende, Roswitha
1994 "Die Provinz Sachsen und das Herzogtum Anhalt im Kaiserreich." In *Geschichte Sachsen-Anhalts III: Bismarckreich bis Gründung der Bezirke 1952*, pp. 41–80. Landesheimatbund Sachsen-Anhalt e.V. Munich/Berlin: Koehler und Amlang.

Ministerium für Ernährung, Landwirtschaft und Forsten Sachsen-Anhalt
1993 "Buchführungsergebnisse landwirtschaftlicher und gartenbaulicher Betriebe. Wirtschaftsjahr 1992/93." Bearbeiter: Dr Reiner Hartung, Peter Zack: Lehr-und Versuchsanstalt des Landes Sachsen-Anhalt, Bereich Acker-und Pflanzenbau, Seminar für Beratung und Informationswesen, Bernburg.

Nagengast, C.
1991 *Reluctant Socialists, Rural Entrepreneurs: Class, Culture, and the Polish State*. Boulder: Westview.

Narayan, K.
1993 "How Native is a 'Native' Anthropologist?" *American Anthropologist*, *95*, 671–686.

Nash, June
1987 "Corporate Hegemony and Industrial Restructuring in a New England Industrial City." In *Perspectives in U.S. Marxist Anthropology*, ed. David Hakken & Hanna Lessinger. Boulder: Westview.
1989 *From Tank Town to High Tech: The Clash of Community and Industrial Cycles*. Albany: State University of New York Press.

Neue Landwirtschaft
1990 "Bauernproteste zeigten Wirkung: Einmaliger, umfangreicher Abbau von DDR-Agrarüberschüssen durch Verkauf an UdSSR." *Neue Landwirtschaft*, September.

Paasch, Ernst-Walter
1994 "Solidarität gegen Crashkurs." *Neue Landwirtschaft*, February, 7–9.

Riemann, Alfons
1990 "Felder gemeinsam bewirtschaften—so funktioniert's im Westen." *Top Agrar DDR-Spezial*, September 21, 26–28.

Rühle, G.
1991 "Der Weg der DDR-Landwirtschaft von 1945 bis zur vollständigen Kollektivierung." *Oekologie und Landbau*, *77*, 7–10.

Sahlins, Marshall
1985 *Islands of History*. Chicago: University of Chicago Press.

Schmidt, F.
1996 *Sozioökonomische Tendenzen im Vergleich bäuerlich geprägter Haupt- und Nebenbetriebe aus den neuen und alten Bundesländern*. Halle/Salle: Dissertation.

Schmidt, K.
1992 "Tiefstpunkt noch nicht erreicht: Der Arbeitskräfteabbau, seine Folgen und mögliche Alternativen." *Neue Landwirtschaft*, November, 14–16.

Schremmer, Heinz
1990 "Haben unsere Schweineanlagen Chancen im europäischen Markt?" *Neue Landwirtschaft*, April, 41–43.

Schunter-Kleeman, Susanne
1995 *'Bei der Bäuerin wird gespart.' Aspekte der Lebenssituation von Landfrauen in vier europäischen Ländern.* Berliner Journal für Soziologie, Heft 2.

Szelényi, I.
1988 *Socialist Entrepreneurs: Embourgeoisement in Rural Hungary.* Madison: University of Wisconsin Press.

Thiele, Holgar
1998 *Dekollektivierung und Umstrukturierung des Agrarsektors in den neuen Bundesländer. Agrarwitschaft*, Sonderheft 160. Frankfurt: Agri-Media.

Top Agrar Spezial
1991 "Ostdeutscher Weizen hat Spitzenqualität." November 25, 3.
1991 "30 Prozent weniger Kartoffeln." October 14, 25.
1991 "Viehställe in schlechtem Zustand." November 15, 13.
1991 "Zwei Prozent weniger Milchquote." December 23, 3.
1993 "Einen völligen Altschuldenerlaß wird es nicht geben." November 13, 3.

Tümmler, Edgar, Konrad Merkel, & Georg Blohm
1969 *Die Agrarpolitik in Mitteldeutschland und ihre Auswirkung und Verbrauch landwirtschaftlicher Erzeugnisse.* Wirtschaft und Gesellschaft in Mitteldeutschland. Vol. 3. Berlin: Duncker & Humblot.

Turner, Lowell
1997 "Introduction: Up Against the Fallen Wall: The Crisis of Social Partnership in Unified Germany." In *Negotiating the New Germany: Can Social Partnership Survive?*, ed. L. Turner, pp. 1–14. Ithaca: ILR Press (Cornell University Press).

Vasary, I.
1987 *Beyond the Plan: Social Change in a Hungarian Village.* Boulder: Westview.

Verdery, Katherine
1996 *What was Socialism, and What Comes Next?* Princeton: Princeton University Press.

Visweswaran, K.
1994 *Fictions of Feminist Ethnography.* Minneapolis: University of Minnesota Press.

Voskamp, U., & V. Wittke
1991 "Aus Modernisierungsblockaden werden Abwärtsspiralen—zur Reorganisation von Betrieben und Kombinaten in der ehemaligen DDR." *Berliner Journal für Soziologie, 1.*

Wenkart Smollett, E.
1984 "The Social Transformation of the Bulgarian Countryside." In *Culture and Community in Europe: Essays in Honor of Conrad M. Arensberg*, ed. O. Lynch. Delhi: Hindustan.

Wilke, G.
1986 "The Sins of the Fathers: Village Society and Social Control in the Weimar Republic." In *The German Peasantry: Conflict and Community in Rural Society from the Eighteenth to the Twentieth Centuries*, ed. R. J. Evans & W. R. Lee, pp. 174–204. London: Croom Helm.

Wissing, P., & S. Anz
1993 "Mit Altschulden wirtschaften." *Neue Landwirtschaft*, March.

INDEX